Seed Storage
of Horticultural Crops

S. D. Doijode, PhD

CRC Press
Taylor & Francis Group
Boca Raton   London   New York

CRC Press is an imprint of the
Taylor & Francis Group, an informa business

Reprinted 2010 by CRC Press

CRC Press
6000 Broken Sound Parkway, NW
Suite 300, Boca Raton, FL 33487
270 Madison Avenue
New York, NY 10016
2 Park Square, Milton Park
Abingdon, Oxon OX14 4RN, UK

Published by

Food Products Press®, an imprint of The Haworth Press, Inc., 10 Alice Street, Binghamton, NY 13904-1580.

Cover design by Marylouise E. Doyle.

Library of Congress Cataloging-in-Publication Data

Doijode, S. D.
    Seed storage of horticultural crops / S.D. Doijode.
        p. cm.
    Includes bibliographical references and index.
    ISBN 1-56022-883-0 (hard : alk. paper) — ISBN 1-56022-901-2 (soft : alk. paper).
    1. Horticultural crops—Seeds—Storage. I. Title.

SB118.4 .D65 2001
635'.0421—dc21
                                                                    00-020623

To my loving mother, Tujai

ABOUT THE AUTHOR

S. D. Doijode, PhD, is an eminent horticulturalist who has worked on seed storage for 20 years at the Indian Institute of Horticultural Research (IIHR), Bangalore. He is Head of the Section on Plant Genetic Research. He earned his PhD in Horticulture at the University of Agricultural Sciences, Bangalore, where he now teaches graduate courses.

The author of more than 100 research publications in professional journals, he is also the author of three book chapters. Dr. Doijode is a life member of the Indian Society for Vegetable Sciences. He has made scientific trips to several countries to discuss and share research experiences in seed storage.

CONTENTS

SECTION IV: SEED STORAGE
IN ORNAMENTAL CROPS

Foreword

Fruits and vegetables play a major role in human nutrition, while ornamental plants provide a healthy environment for living. The cultivation of these crops is rapidly expanding in different parts of the globe. Large areas are being brought under cultivation of horticultural crops. Several vegetable and ornamental crops and a few fruit species are commercially propagated by seeds. Seeds are also used in raising rootstocks during propagation, especially in perennial fruit crops. Seeds also contribute greatly in effective conservation of germplasm in the gene bank. Seed viability differs in various horticultural crops, and it is short-lived. Further, it needs suitable storage conditions for maintaining high seed quality for longer periods. High seed quality in terms of viability and vigor contributes to higher crop production. Many times, farmers experience crop failures due to sowing of poor quality seeds obtained from improper storage.

This book appears to be the first of its kind on the storage of horticultural crop seeds. I am pleased to note that Dr. Doijode has utilized his two decades of experience in seed storage, particularly in horticultural crops, to produce this publication.

The book contains the latest information on various aspects of seed storage in different fruits, vegetables, and ornamental crops of temperate, tropical, and subtropical regions, emphasizing the latest concepts, developments, and approaches for effective seed conservation. It is hoped that this book will be greatly beneficial to students, researchers, teachers, horticulturalists, the seed industry, and, above all, the farming community, for appreciating the importance of developing effective seed conservation strategies for various horticultural crops.

K. L. Chadha, PhD, DSc, FASS
ICAR National Professor (Horticulture)
Indian Agricultural Research Institute
New Delhi, India

Preface

Seed is a basic input in horticulture. Different horticultural crops are valued for their delicious fruits, enriched with vitamins and minerals, and appreciated for their ornamental qualities. Seeds are sown in many vegetable crops, a few fruit crops, and ornamental crops. However, they are predominantly used in raising rootstocks for vegetative propagation in different perennial fruit crops. Use of high-quality seeds in terms of viability and vigor imparts higher yield and better-quality produce. Also, seeds contribute to highly valued genetic material needed for crop improvement, production, and conservation. At present, seeds are the only viable medium that is comparatively stable during long-term conservation. Seeds are normally stored for short periods, such as until the next growing season, but in special circumstances, such as genetic conservation, they are stored for longer periods. Improper seed storage reduces seed viability and causes great losses within the farming community.

Farmers, nurserymen, and breeders face many hardships in maintaining high seed quality during storage. My two-decades-long association with seed storage in horticultural crops helped produce this volume, which gathers together the scattered information on seed storage behavior and means of storage for various fruits, vegetables, and ornamental crops of tropical, subtropical, and temperate regions.

The book will be useful for graduate and postgraduate students and teachers of horticulture, seed science and technology, and plant genetic resources faculties. It also imparts valuable scientific information to researchers and the seed industry on the preservation of valuable planting material.

I express my deep gratitude to Dr. Amarjit S. Basra for the initiation of this work, as well as my appreciation for his encouragement, and to The Haworth Press for readily accepting the manuscript and for the excellent end product. I am particularly indebted to Bill Palmer, Patricia Brown, Peg Marr, Dawn M. Krisko, Amy Rentner, Melissa Devendorf, and other staff at The Haworth Press for their encouragement and assistance during the preparation of this manuscript. Without their active support the book could not have been completed.

I am extremely grateful to Dr. K. L. Chadha, ICAR National Professor (Horticulture), Indian Agricultural Research Institute, New Delhi, for writing the foreword to this book.

I am grateful to Dr. I. S. Yadav, Director, Indian Institute of Horticultural Research, Bangalore, for support and encouragement. I am thankful to several of my friends, particularly Mr. P. N. Krishnamurthy, for their valuable help in completing this book.

I am indebted to my parents, Shri Dattatraya and Smt Kantabai, who taught me gardening and inspired me to study horticulture.

My sincere appreciation goes to my family, Vasant, Gayatri, and particularly to my wife, Jayashree, and children, Amit Kumar, Suppriya, and Rashmi, for their inspiration, encouragement, and support in completing this book.

S. D. Doijode

SECTION I:
INTRODUCTION

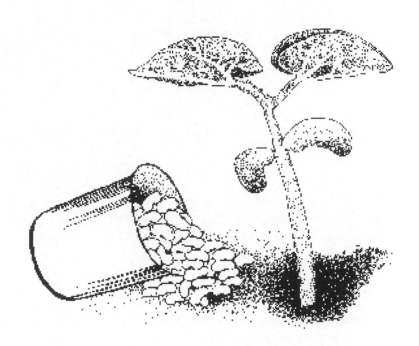

Seeds and Seed Storage

IMPORTANCE OF SEEDS IN HORTICULTURE

The viable seed is a source of a new plant, a beginning based on inherited parental characteristics. Seeds contain genetic material in compact form that is well protected from extraneous factors. Different horticultural crops, such as fruits, vegetables, and ornamental crops, are propagated by either sexual or asexual reproduction, or by both means. Normally, seeds multiply annuals, whereas vegetative propagation is used in perennial crops. Even in the latter, seeds are used for raising rootstocks and also in breeding programs. Seeds are a viable tool for long-term conservation of genetic diversity because they are easier to handle, practicable, inexpensive, and capable of maintaining genetic stability on storage.

Several horticultural crops contribute to the welfare of people and the economies of nations. Many of them are eaten directly or preserved and used as processed food. These are rich sources of proteins, carbohydrates, vitamins, minerals, and fiber and are low in fats and oils. Fruits and vegetables greatly contribute to a nutritional and healthy diet. Ornamental crops provide pleasing and beautiful landscaping that contributes to people's mental health.

Seeds are largely involved in evolutionary processes in the plant industry. Vast genetic diversity is the result of repeated natural propagation by seeds. The farming community also depends on quality seeds for rich harvests. In general, seeds have diverse uses. They provide food, are a raw material, and protect human life. Seeds are a source of proteins, carbohydrates, and oils for the staple human diet. They also are a source of spices, condiments, and medicines. Seeds are used as well in the manufacturing of gum, cosmetics, and paints.

In botanical terms, a seed is a fertilized ovule, and for practical purposes, it is a dry unit of propagation that transmits genetic material from generation to generation. A seed consists of a tiny embryonic plant in the resting stage that is provided with food and well protected by the hard seed covering. A seedling emerges from the seed upon germination under favorable conditions. Seeds vary in size, shape, and color. They can be oval, round, triangular, cylindrical, and sometimes irregular in shape. Their surface can be rough, smooth, or covered with hairs. Colors include red, yellow, green, white, black, or brown in various crop plants.

SEED DEVELOPMENT

Seeds are formed inside the botanical fruit, which as a mature ovary contains one or more ovules that develop into true seeds. Seeds form because of double fertilization. The megaspore diploid mother cell gives four haploid cells on meiosis during megasporogenesis. Normally, one megaspore is functional and carries the chromosomes of the mother plant, while the other three degenerate. This single cell enlarges and becomes an embryo sac. The cell undergoes three nuclear divisions to form eight haploid nuclei. Three nuclei will arrange as antipodal cells at one end, two polar nuclei will locate near the center, and three will gather at the other end to form an egg and two synergid cells. The polar nuclei fuse and form a diploid nucleus. During pollination, pollen germinates and the pollen tube enlarges and grows into the embryo sac. The tube nucleus degenerates and the two pollen nuclei enter the embryo sac; one fuses with the diploid polar nucleus to form an endosperm, and another fuses with the egg cell to form a zygote. Cell division occurs rapidly after syngamy to form an embryo (see Figure 1.1). In dicotyledonous plants, the endosperm formed is used during seed development. This seed is mainly composed of embryo. Similarly, the endosperm degenerates in orchid seeds. In monocotyledonous seeds, the endosperm develops and stores the food material required during germination. There is a gradual increase in seed weight during development due to cell growth and elongation. Initially, the seed moisture content is high, but it decreases with growing maturity. Simultaneously, the dry weight increases and stabilizes during the mass maturity. The seed contains carbohydrates, proteins, and oils and supplies the food material to the young seedling during seed germination.

SEED GERMINATION AND FACTORS
AFFECTING SEED GERMINATION

The resting embryo emerges as a young plant upon germination under favorable conditions (see Figure 1.2). The embryo absorbs moisture, activates the metabolism, and grows; the young seedling emerges after rupturing the seed coat. In many cases, fresh seeds germinate readily, whereas others re-

FIGURE 1.1. Longitudinal Section of Flower and Double Fertilization

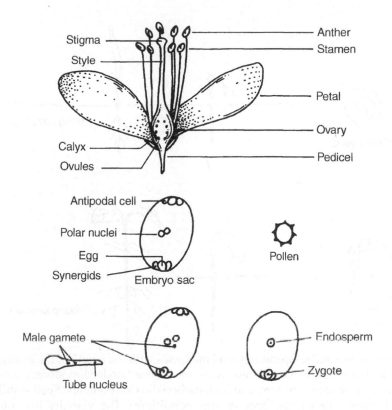

quire after-ripening to eliminate the dormancy. Certain changes occur during after-ripening that facilitate the germination process. On imbibition, the cell becomes more turgid and facilitates the exchange of gases. This activates the enzyme system that breaks the stored food material and translocates it to the growing point. Normally, the radicle emerges first by rupturing the seed coat; later, both root and shoot systems develop, and the young seedling becomes independent and synthesizes its own food material for growth and development.

Initial Viability

Seed begins to deteriorate on separation from the mother plant. Seed deterioration is promoted under unfavorable storage conditions, such as high

FIGURE 1.2. Seed Germination in Monocotyledonous and Dicotyledonous Seeds

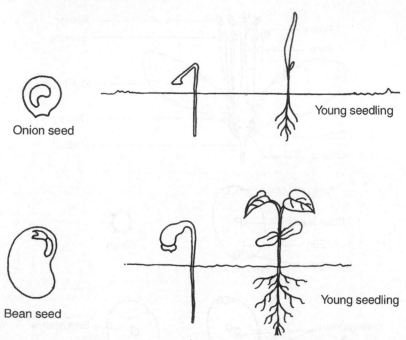

seed moisture and temperatures, and increases with the length of the storage period. However, metabolic processes are reduced under low storage temperature, low seed moisture, and high carbon dioxide contents. Seed viability decreases under improper storage conditions. The viability loss rate varies in different species. In certain species seeds are viable for a very short period. Fresh and nondormant seeds exhibit higher germination, although it subsequently declines with improper storage.

Seed Dormancy

Seed dormancy normally refers to the failure of viable seeds to germinate under favorable germination conditions. Presence of dormancy is beneficial to the extent that it prevents sprouting during open storage or in rains during harvesting. Most seeds are dormant at the time of harvest. This dormancy is due to improper development of the embryo, a seed coat impermeable to water and gases, and the presence of inhibitors. The seed coat acts as a barrier to water, oxygen uptake, and radicle emergence. It also contains inhibitors and sometimes carries pathogens affecting the germination process.

Softening of the seed coat occurs over time in different species due to hydration and dehydration, freezing and thawing, soil acidity, and attack by microorganisms. Seed scarification, physically or chemically, softens the seed coat, facilitating easy absorption of moisture and exchange of gases. Hard seeds are commonly found in species of Leguminosae, Compositae, and Malvaceae. Legume seed hardness develops on exposure to low humidity or very dry conditions. In cases of embryo dormancy, seeds fail to germinate even after removal of the seed coat. Such seeds are commonly found in temperate fruits and require cold stratification for germination. Most of these seeds are stored at 3 to 10°C for various periods to eliminate the dormancy. Embryo dormancy lasts for various periods in different genotypes. Sometimes seeds possess both a seed coat and innate dormancy. In beetroot seeds, germination inhibitors present in the seed coat prevent the germination process. Chemicals, such as hydrogen cyanide, ammonia, and ethylene inhibit germination. The naturally occurring inhibitors are cyanide, ammonia, alkaloids, organic acids, essential oils, and phenolic compounds. In enforced dormancy seeds fail to germinate due to nonavailability of parameters, such as absence of light or low temperatures. Such seeds germinate under suitable light or temperature conditions.

Dormancy Removal Methods

Seed dormancy is genetically controlled and modified by pre- and postharvest environmental factors. Use of certain treatments is beneficial in overcoming dormancy in particular crop plants.

After-ripening. Seed germination improves with increase in storage period. Certain changes occur in dry storage conditions that result in higher seed germination. The after-ripening period is shorter under higher storage temperatures and higher seed moisture. However, under similar circumstances, seed viability is lost rapidly. Normally, seeds are dried before storage. Such predrying of seeds also serves as after-ripening.

Scarification. The most common method to reduce seed hardness is the scarification of testa. A hard seed coat restricts the movement of moisture vapor and gases. Seeds are scarified by mixing them with sharp sand and shaking vigorously for 10 to 20 minutes (min). This damages the hard seed coat. For smaller seed quantities, the seed coat is removed or scarified by soaking in acids. Seeds are soaked in concentrated sulfuric acid up to 20 min depending on seed hardness. In certain cases, a longer duration of soaking affects seed germination, resulting in the production of abnormal seedlings. Seed soaking in ethyl alcohol dissolves the waxy cuticle and promotes germination. Seed soaking in boiling water also ends dormancy, but it injures the seeds in several instances. The seed coat can also be softened with higher temperatures (50 to 60°C) or by soaking it in liquid nitrogen for 5 min. However, improper seed handling causes loss of viability.

Stratification. Stratification refers to exposure of moist seeds to chilling temperatures. These stratified seeds germinate slowly at lower temperatures and rapidly at higher temperatures. In certain cases, warm and cold stratification promotes seed germination. Exposing *Prunus* seeds to 25°C for two weeks, followed by exposure to low temperature, eliminates the dormancy.

Light

Light influences the germination of dormant seeds. In certain crop seeds, light either promotes or inhibits the germination process. Photochemical reactions occur between 290 to 800 nanometers (nm) radiation. The molecule absorbs the light particles, which stimulates the germination process. In general, wavelengths of 660 to 700 nm (red light) promote germination, and wavelengths of 700 to 760 nm (far-red light) are inhibitory. The main pigment involved in germination is phytochrome (Thanos and Mitrakos, 1979), a protein molecule that occurs in two forms, red (Pr) and far red (Pfr). Pr absorbs mainly red light, with a peak absorption at 670 nm, and Pfr absorbs far-red light, with a peak absorption at 720 to 730 nm.

Photoperiod also affects germination. Some seeds respond well to short-day conditions and others to long-day conditions. The presence of continuous light induces dormancy and inhibits germination.

Seed Washing

Germination inhibitors are present on seed or fruit coats. Thus, soaking the seeds and changing the water can improve germination. Repeated washing of seeds softens the tissues and facilitates early seedling emergence.

Chemicals

Several chemicals, either dissolved in acetone or in water, promote the germination process. The most commonly used chemicals are hydrogen peroxide, ethyl alcohol, thiourea, sodium hypochlorite, potassium nitrates, gibberellins, and cytokinins. Gibberellic acid substitutes for the light requirements and promotes seed germination.

SEED STORAGE

Seed storage preserves seed viability and vigor for various periods by reducing the rate of seed deterioration. Seed storage is essential for preserving high seed quality, vigor, and viability for future use by farmers or breeders.

The latter requires a fairly long storage period to avoid frequent growing of the crop as well as to maintain genetic variability for future use. Longer seed storage is achieved by preserving seeds in climatic conditions favorable for storage and/or by modifying the environment around the seeds so that it is suitable for longer storage. Seed quality is maintained, but does not improve, during storage. Seeds commence deterioration upon separation from the mother plant. It is necessary to prevent deterioration by providing suitable storage conditions. Low temperature and low seed moisture are the two most effective means of maintaining seed quality in storage. Most of the vegetable seeds store well for eight to ten months at 8 to 10 percent moisture content (mc) and 20°C (Delouche, 1990). Dry, cool conditions are best for longer storage. Roberts (1983) broadly classified seeds as orthodox and recalcitrant. Orthodox seeds can be dried to low moisture content and stored longer at low temperatures, whereas recalcitrant seeds survive for a few weeks or months in storage and are killed by desiccation below a certain high moisture content, e.g., mango, mangosteen, rambutan, jackfruit, durian, and avocado. The recalcitrant seeds present complex problems. They normally perish in a week (Chin, 1978) but can survive for one to six months under favorable storage conditions. In certain species, storage behavior is intermediary. Several factors contribute to longer storage life. Brett (1952) noted that high-quality, sound seed ensures longer retention of maximum viability during storage. Barton (1941) reported that seeds of high initial viability are more suitable for storage. Bass (1980) suggested the involvement of other factors, such as growing conditions, maturity at harvest, and harvesting and processing conditions. Correct seed moisture content and storage temperatures are also vital for proper seed storage.

SEED VIGOR

Seed vigor plays an important role during storage of seeds. Vigorous seeds possess greater storage potential and preserve well for longer periods until seeds become nonviable. Vigor is a relative term used to estimate seed quality. Vigorous seed produces strong and healthy uniform seedlings having better field establishment and exhibiting relatively greater longevity. The viability trend reduces gradually during storage, followed by a sharp decline. Similarly, seed vigor reduces and is parallel to the viability trend, although at lower levels. Seed vigor is a characteristic of a genotype that improves with progress in maturity level. Fully matured seeds, which have completed seed development, show higher seed vigor. Bold and high-density seeds also show more vigor.

Improper seed development, stressful growing conditions, and inadequate nutrients affect seed vigor. Such low-vigor seed reduces growth rate and yields more abnormal seedlings, delays maturity, and decreases crop yield. Seed vigor also is reduced during aging of seeds, and by insect and fruit infestations, although the process of seed aging slows down under favorable storage conditions. Seed vigor is generally determined by seedling growth, speed of germination, measurement of seed leachates, oxygen consumption and carbon dioxide released, tetrazolium, glutanic acid decarboxylase activity (GADA) test, and different types of stress tests. Seed vigor improves especially in medium- and low-vigor seed lots. Midstorage soaking of seeds for 2 to 5 hours (h) or in water-saturated atmosphere for 24 to 48 h, followed by thorough drying, improves vigor in the majority of orthodox seeds.

VARIETAL DIFFERENCES

Seed longevity is primarily dependent on genetic factors. Thus, it differs among genera and species, and sometimes among genotypes. Hard seeds, especially those in the Leguminosae family, live longer, whereas seeds of onion, lettuce, papaya, and recalcitrant seeds have shorter storage lives. James, Bass, and Clark (1967) reported significant variations in storage of different cultivars of tomato, bean, cucumber, pea, and watermelon. Similarly, varietal differences were observed in bean (Toole and Toole, 1954) and in lettuce (Harrison, 1966). The information on inheritance of longevity is scanty. Crosses of inbred lines with shorter storability resulted in short-lived seeds, while parents with longer storability produced hybrid seeds with greater longevity (Haber, 1950).

PREHARVEST FACTORS

Environmental factors such as temperature, rainfall, humidity, nutrition, and irrigation are liable to change the seed characteristics during seed development and maturation. Deficiencies of elements such as nitrogen, potassium, and calcium significantly affect the storability of seed lots (Copeland, 1976). Seed size also affects the storage behavior of seeds. Large to medium-sized seeds retain viability for longer periods. Mackay and Tonkin (1967) recorded a positive correlation between weather conditions during ripening and harvesting of seeds. Haferkamp, Smith, and Nilan (1953) showed that healthy, mature seeds harvested in dry weather are ideal for longer storage. Harrington and Thompson (1952) reported that region of seed production affects seed viability.

Seed Maturity

Seed attains maximum dry weight upon maturity (Harrington, 1972). At the same time, it attains maximum quality at the end of the seed-filling phase (mass maturity); thereafter, seed viability and vigor decline (Ellis, Demir, and Pieta, 1993). In certain vegetable crops maximum potential longevity was not attained until after mass maturity. Carrot seeds mature over a period of time, and maturity affects seed viability and seedling vigor. Mature and healthy seeds remain viable longer than immature seeds. Rapid loss of viability is observed in immature seeds of cabbage, cucumber, eggplant, pumpkin, tomato, and watermelon as compared to mature seeds (Eguchi and Yamada, 1958).

STORAGE CONDITIONS

Seed Moisture

Seed moisture plays a major role in seed storage. Higher seed moisture is injurious to seed life and rapidly reduces viability during storage. Harrington (1972) suggested that seed longevity decreases by one-half for every 1 percent rise in seed moisture, and that it maintains well at the moisture range of 5 to 14 percent. Seed moisture varies in different species as a result of differences in chemical composition of seeds and conditions prevailing during seed development and maturation. Seeds of particular species have their own moisture holding capacities and absorbing potentials. High-moisture seeds are more susceptible to heat damage. The biological activity, growth, and multiplication of insects and pathogens are greater at higher levels of moisture. The rate of seed deterioration enhances with increased moisture levels. Seeds of onion, eggplant, and pepper were preserved at 10 to 100 percent relative humidity (RH) at 20°C. Seed viability was rapidly reduced at higher humidity (40 percent). High seed viability was maintained at 10 percent RH for four years (Doijode, unpublished data). Seeds absorb moisture from the atmosphere due to a gradient created in vapor pressure, whereby the moisture moves from high to low vapor pressure. The rate of absorption depends on the chemical composition of the seed. Seeds with high protein content absorb moisture more rapidly than those with high levels of carbohydrates and fats. In general, the moisture content of seed increases with a rise in relative humidity level, and absorption also depends on the temperature. The water holding capacity of air increases with an increase in atmospheric temperature. Normally, absorption is greater for the first two or three days of storage. It will reach equilibrium with atmospheric humidity at a given time during storage. Seed moisture content fluctuates with long-term changes in atmospheric humidity.

Seed moisture of 5 to 7 percent is ideal for safe storage of seeds. A moisture gradient develops during drying, and internal moisture moves toward the surface, from which it evaporates. The embryo is damaged by rapid removal of seed moisture. Water present in vapor form in intercellular space is removed by evaporation. Seeds are dried by several means, such as natural drying, sun drying, dehumidified drying, drying with desiccants, vacuum and freeze drying. Cromarty, Ellis, and Roberts (1982) recommended seed drying at low temperature (15°C) and low humidity (15 percent RH) for safe storage. Zhang and Zhang (1994) reported that ultradrying of seeds improves seed storability. Ultradry (2.0 to 3.7 percent mc) seeds preserve viability better than dry (5.5 to 6.8 percent mc) seeds at 20°C (Ellis et al., 1996). Insect activity stops at moisture equal to <35 percent RH; similarly, fungal activity is inhibited at less than 65 to 70 percent RH. Seed moisture increases especially in sealed storage; therefore, well-dried and low-moisture seeds are preserved in moisture-proof containers during storage.

Storage Temperature

Storage temperatures, in combination with moisture, play an important role in seed deterioration. Seed longevity in several horticultural species decreases with an increase in storage temperature. Harrington (1972) reported that seed life decreases by one-half when temperatures are increased by 5°C in the temperature range of 0 to 50°C. Seeds of several species tolerate freezing temperatures, especially at low levels of seed moisture, thereby enhancing seed longevity. Roberts (1972) reported that low-temperature storage is advantageous for maintenance of seed viability, with few exceptions. Seed storage at fluctuating temperatures damages seed quality (McRostie, 1939). Horky (1991) noted that seeds of certain vegetable crops having 5.6 to 7.5 percent mc maintained their initial viability for seven years at 0°C, while high moisture (22 percent mc) causes injury to seeds when stored at −6°C (Agena, 1961). Low temperature is effective in preserving seed quality, especially in moisture-proof containers under humid conditions. Seeds of tomato, cucumber, and sweet pepper are viable for 36 months at 20°C and for 70 months at 0°C (Zhang and Kong, 1996). Stanwood and Sowa (1995) preserved different types of seeds at −18°C and −196°C. Seed viability did not decline over the ten years. The percentage of germination of seeds stored at 5°C dropped from 94 to 68 percent, which is attributed to the variation in moisture content. It is less likely that the variation in seed moisture content was a factor in the −18°C storage. Seed deterioration is greater at 5°C than at −18 and −196°C.

Oxygen

Higher oxygen content during storage reduces seed viability (Roberts and Abdalla, 1968). The response is inconsistent and varies in different species. The deleterious effects of oxygen are more pronounced at higher seed moisture levels. Seeds stored in oxygen lose viability sooner than those stored in carbon dioxide and nitrogen (Sayre, 1940). Barton (1960) reported that seeds sealed in oxygen lost viability rapidly and that storage in carbon dioxide, nitrogen, or partial vacuum extended storage life for six to eight years. In contrast, Bass, Clark, and James (1963) saw no significant advantage in using partial vacuum, carbon dioxide or nitrogen, argon or helium instead of airs. Longevity of pea seeds decreases with an increase in oxygen concentration from 0 to 21 percent (Roberts and Abdalla, 1968). Harrison (1956) reported that seed stored with carbon dioxide retains better viability than seed sealed in air. Tao (1992) observed that the number of crops reported to have vacuum storage is rather small. High seed moisture and warmer storage adversely affect the viability of vacuum-stored seeds. Nitrogen diffuses slowly into the seeds, delays the loss of viability, and inhibits insect activity (Yadav and Mookherjee, 1975) and mold growth (Glass et al., 1959). More research is needed to harness the benefits of gaseous storage of seeds.

Seed Packaging

Seeds are packed in different containers to protect them from extraneous environmental factors, to facilitate easier handling during storage, and to enhance their marketability. Seeds should be suitably protected from high atmospheric humidity, and also from storage insects and pathogens. Normally, seeds are packed in cotton, jute, paper bags, polyethylene bags, laminated aluminum foil pouches, or aluminum cans. The choice of container depends on the quantity to be stored, the duration of storage, the cost of seed material, and the storage conditions. In open storage, humidity is controlled using a dehumidifier, while at low temperatures, moisture-proof containers are used to protect seeds from high humidity. Normally, hermetic containers such as laminated aluminum foil pouches or cans are used for long-term conservation. These pouches are effective in maintaining desired levels of seed moisture for fairly long periods. Low-moisture seeds should be packed in sealed packages.

Laminated pouches are handy and attractive, can be cut to various sizes, are reusable, can withstand temperatures from -20 to $40°C$, and occupy less space during storage. Earlier, Duvel (1905) reported that seeds in unsealed containers deteriorated rapidly under high temperatures and humidity, whereas those in sealed containers maintained reasonably high viability. Seeds of many vegetable crops retained viability when stored in tin cans, and containers of chlorinated rubber, waxed paper, cellophane, and paper provided superior storage (Coleman and Peel, 1952). In sealed containers, dry seeds with 5 to 7 percent mc retained high germination energy both at low (0 to $5°C$ or $-15°C$) and at high ($25°C$) temperatures (Sokol and Kononkov,

1975). Sometimes volatile substances such as formaldehyde emerge from the printing ink, paper, or glue on seed packets, which causes rapid decreases in seed germination. These toxic compounds are more effective at higher seed moisture (Kraak and Pijlen, 1993). It is essential to avoid use of such containers, especially during long-term conservation.

Storage Pests

Unhygienic and poor sanitary conditions during storage often lead to the presence of pests, such as insects, fungi, and viruses, which are commonly found on seeds. These storage pests are more active at higher seed moisture and higher temperatures. *Alternaria, Aspergillus, Chaetomium, Cladosporium, Fusarium, Helminthosporium,* and *Penicillium* fungi associate with seeds. *Aspergillus* and *Penicillium* are common storage fungi that even grow under dry conditions. Seed infestations with fungi result in loss of viability, reduction in vigor, discoloration, and generation of heat and rotting. Storage fungi grow well between 4 to 45°C and 65 to 100 percent RH. Seed storage in carbon dioxide and nitrogen inhibits the growth of microorganisms. Saprophytes such as *Aspergillus, Penicillium,* and *Rhizopus* are effectively controlled by applications of Deltan, thiram, and mancozeb (Gupta and Singh, 1990). Seed treatment with organic mercury compounds greatly reduces eggplant viability and causes injury to cabbage and beet seeds (Nakamura, Sato, and Mine, 1972).

Insects commonly found in storage are the weevil, grain borer, grain moth, and beetle. Weevils destroy the endosperm and attack the embryo, thereby reducing germination capacity. Insect activity is reduced under clean conditions and decreasing moisture. Fumigation with proper dosage of pesticides controls insects and does not impair seed quality.

SEED DETERIORATION

Deterioration is a natural phenomenon in living material. The rates of deterioration in seeds vary in different species and depend on several factors, including storage conditions. Seed deterioration during storage involves progressive impairment of performance and function of seeds. It causes reduced seed quality, viability, and vigor, leading finally to seed death. Seed deterioration is an irreversible process; it cannot be prevented or reversed, but it can be slowed under specific conditions. Seed deterioration is greater under improper storage conditions, such as higher seed moisture and temperatures. Evidence of seed deterioration includes discoloration of seeds, poor germination, poor seedling growth, production of more abnormal seedlings, and poor yield. Physiological changes noticed in seeds are reduced respiration, loss of enzyme activity, and membrane damage. The exact mechanism of seed deterioration is unclear; however, different workers have suggested several possibilities. Roberts (1981) reported that rate of aging depends on three main factors, namely, temperature, moisture, and oxygen pressure. Recalcitrant seeds are killed by low moisture and can be stored for a relatively short period under moist conditions, whereas orthodox seeds store fairly well under dry, cool conditions. Some damages caused during storage are reversible under hydrated conditions, probably through the cellular repair mechanism. Severely damaged seeds exhibit low vigor and produce more abnormal seedlings.

Beweley (1986) noted that there are no clear-cut answers for the membrane deterioration in stored seeds. Lipid degradation is a major cause of deterioration, releasing free fatty acids, which initiates oxidative deterioration by providing substrate for lipoxygenase, which primarily attacks membranes. Low-vigor embryos contain higher levels of free radicals than high-vigor ones. Further application of vitamins C and E reduces oxidative lipid damage and prevents deterioration to some extent (Sathiyamoorthy and Nakamura, 1995). Also, Vertucci (1992) noticed changes in lipid constituents of seeds in relation to loss of seed viability during storage. Trawartha, Tekrony, and Hildebrand (1995) also suggested that lipid peroxidation and generation of free radicals contribute to seed deterioration. Lipoxygenase (LOX) is thought to be a major contributor to lipid peroxidation and free-radical generation. Decrease in fatty acids content was observed in broad bean and carrot, but not in cucumber, pea, squash, capsicum, and onion, during the aging process (Perl, Yaniv, and Feder, 1987). Beral-Lugo and Leopold (1992) reported that soluble sugars are important components of proteins and contribute to the deterioration changes occurring during seed storage. Reduction in seed vigor is associated with a decline in monosaccharides, while raffinose and sucrose contents remain stable. Soluble carbohydrates play an important role in desiccation tolerance and seed storability (Steadman, Pritchard, and Dey, 1996). Solute leakage occurs more in deteriorated seeds due to membrane damage (Schoettle and Leopold, 1984).

The reduction in viability is also due to the evolution of volatile compounds such as methanol, acetaldehyde, ethanol, and acetone during storage. The harmful effects are increased with longer storage periods, and ethanol deteriorates at higher levels of RH. Zhang and colleagues (1994) reported that acetaldehyde, the most deleterious volatile compound, accumulates even at $-3.5°C$. It is suggested that endogenous volatile compounds such as acetaldehyde accelerate seed deterioration at low humidity and low temperatures throughout long-term storage.

Chromosomal changes such as ring formation, bridges, and fragmentation occur in chromosomes upon aging of seeds. Seeds stored under proper conditions do not show chromosomal aberrations (James, 1967) and give yields similar to those of fresh seeds. Seeds stored at −5°C for 13 years show unaffected yields similar to those of fresh seeds (Barton and Garman, 1946). Yield was reduced in deteriorated onion seeds as compared to fresh seeds (Newhall and Hoff, 1960).

SEED INVIGORATION

High-vigor seed stored better over a period. Seed invigoration improves the vigor in medium- and low-vigor seed lots. Villiers and Edgcumbe (1975) suggested that a genetic maintenance and repair system is feasible under imbibitional storage conditions, but these conditions prevent germination, as with seeds buried in soil that remain viable without germination. Roberts (1981) reported that some damages during deterioration are reversible under hydrated conditions and can undergo repair, whereas severe damage in seeds is irreversible. Seed soaking in water for 2 to 5 h or in water-saturated atmosphere for 24 to 48 h, followed by drying, resulted in higher seed viability and longevity. Prestorage applications of antiaging chemicals such as iodine, chlorine, bromine, methanol, ethanol, and propanol improve seed storability. Iodination is effective in maintaining high viability and vigor (Basu and Rudrapal, 1979). The mode of invigoration action is unclear. It prevents damaging oxidative reactions, especially free-radical-induced lipid peroxidation reactions, and repairs the cellular system involved in invigoration (Basu, 1990). Antioxidants that are nontoxic are promising for safe seed storage (Bahler and Parrish, 1981). Butylated hydroxyanisole is particularly effective in retarding deterioration at high temperatures (Okundaye, 1977).

REFERENCES

Agena, M.U. 1961. Untersuchungen Über Kalteeinwirkungen auf lagernde. *Getriedefruchte mit Verschiedenem Wassergehalt Bonn,* p. 112.

Bahler, C.C. and Parrish, D.J. 1981. Use of lipid antioxidants to prolong storage life of soybean seeds. *Am. Soc. Agron.* 5:117.

Barton, L.V. 1941. Relation of certain air temperatures and humidities to viability of seeds. *Boyce Thompson Inst. Contrib.* 12:85-102.

Barton, L.V. 1960. Storage of seeds of *Lobelia cardinalis* L. *Boyce Thompson Inst. Contrib.* 20:395-401.

Barton, L.V. and Garman, H.R. 1946. Effect of age and storage condition of seeds on the yields of certain plants. *Boyce Thompson Inst. Contrib.* 14:243-255.

Bass, L.N. 1980. Seed viability during long-term storage. *Hort. Rev.* 2:117-141.

Bass, L.N., Clark, D.C., and James, E. 1963. Vacuum and inert gas storage of crimson clover and sorghum seeds. *Crop Sci.* 3:425-428.

Basu, R.N. 1990. Seed invigoration for extended storability. *Inter. Conf. Seed Sci. Technol. New Delhi,* Abstr. No. 2.11, p. 38.

Basu, R.N. and Rudrapal, A.B. 1979. Iodine treatment of seed for the maintenance of vigor and viability. *Seed Res.* 7:80-82.

Beral-Lugo, I. and Leopold, A.C. 1992. Changes in soluble carbohydrates during seed storage. *Pl. Physiol.* 98:1207-1210.

Beweley, J.D. 1986. Membrane changes in seeds as related to germination and the perturbation resulting from deterioration in storage. *Spl. Publ. CSSA* 11:27-45.

Brett, C.C. 1952. The influence of storage conditions upon the longevity of seeds with special reference to those of root and vegetable crops. *13th Int. Hort. Cong.,* p. 14.

Chin, H.F. 1978. Production and storage of recalcitrant seeds in the tropics: Seed problems. *Acta Hort.* 83:17-21.

Coleman, F.B. and Peel, A.C. 1952. Storage of seeds. *Queensland Agr. J.* 74:265-276.

Copeland, L.V. 1976. *Principles of seed science and technology.* Minneapolis, Minnesota: Burgess Publishing Company.

Cromarty, A.S., Ellis, R.H., and Roberts, E.H. 1982. *The design of seed storage facilities for genetic conservation.* Rome: International Board for Plant Genetic Resources, p. 96.

Delouche, J.C. 1990. Precepts of seed storage. *Proc. Inter. Sat. Symp. Seed Sci. Technol Hisar* :71-90.

Duvel, J.W.T. 1905. The effect of climatic conditions on the vitality of seeds. *U.S. Dept. Agr. Off. Exp. Stn. Rec.* 16:618.

Eguchi, T. and Yamada, H. 1958. Studies on the effect of maturity on longevity in vegetable seeds. *Natl. Inst. Agr. Sci. Bul. Ser. E. Hort.* 7:145-165.

Ellis, R.H., Demir, I., and Pieta, F.C. 1993. Changes in seed quality during seed development in contrasting crops. *Proc. 4th Inter. Workshop on Seeds, Angers, France* 3:897-904.

Ellis, R.H., Hong, T.D., Astley, D., Pinnegar, A.E., and Kraak, H.L. 1996. Survival of dry and ultra dry seeds of carrot, groundnut, lettuce, oil seed rape and onion during five years of hermetic storage at two temperatures. *Seed Sci. Technol.* 24:347-358.

Glass, R.L., Ponte, J.G., Jr., Christensen, C.M., and Geddes, W.F. 1959. Grain storage studies XXVIII. The influence of temperature and moisture level on the behavior of wheat stored in air or nitrogen. *Cereal Chem.* 36:341-356.

Gupta, A. and Singh, D. 1990. Effect of fungicidal treatments on seed viability and mycoflora of stored vegetable seeds. *Inter. Conf. Seed Sci. Technol. New Delhi,* Abstr. No. 2.20, p. 42.

Haber, E.S. 1950. Longevity of the seeds of sweet corn inbred and hybrids. *Proc. Am. Soc. Hort. Sci.* 55:410-412.

Haferkamp, M.E., Smith, L., and Nilan, R.A. 1953. Studies on aged seeds. I. Relation of age of seed to germination and longevity. *Agron. J.* 45:434-437.

Harrington, J.F. 1972. Seed storage and longevity. In Kozlowski, T.T. (Ed.), *Seed biology,* Volume 3. New York: Academic Press, pp. 145-245.

Harrington, J.F. and Thompson, R.C. 1952. Effect of variety and area of production on subsequent germination of lettuce seed at high temperature. *Proc. Am. Soc. Hort. Sci.* 59:445-450.

Harrison, B.J. 1956. Seed storage. *John Innes Hort. Inst. Ann. Rep.* 46:15-16.

Harrison, B.J. 1966. Seed deterioration in relation to storage conditions and its influence upon germination, chromosomal damage and plant performance. *Nat. Inst. Agric. Bot. J.* 10:644-663.

Horky, J. 1991. The effect of temperatures and long-term storage of dry seeds in some species of vegetables. *Zahradnictvi-UVTIZ* 18:29-33.

James, E. 1967. Preservation of seed stocks. *Adv. in Agron.* 19:87-106.

James, E., Bass, L.N., and Clark, D.C. 1967. Varietal differences for longevity of vegetable seeds and their response to various storage conditions. *Proc. Am. Soc. Hort. Sci.* 91:521-528.

Kraak, H.L. and Pijlen, J.G. 1993. Packaging materials and seed viability, effect of formaldehyde and other toxic volatiles. *Seed Sci. Technol.* 21:463-474.

Mackay, D.B. and Tonkin, J.H.B. 1967. Investigation in crop seed longevity. I. Analysis of long term experiments with special reference to the influence of species, cultivars, provenance and season. *Natl. Inst. Agr. Bot. J.* 11:209-225.

McRostie, G.P. 1939. The thermal death point of corn from low temperatures. *Sci. Agr.* 19:687-689.

Nakamura, S., Sato, T., and Mine, T. 1972. Storage of vegetable seeds dressed with fungicidal dusts. *Proc. Inter. Seed Test. Assoc.* 37:961-968.

Newhall, A.G. and Hoff, J.K. 1960. Viability and vigor of 22 year old onion seeds. *Seed World* 86:4-5.

Okundaye, J.E. 1977. The use of antioxidants as a means of retarding deterioration in high lysine corn and soybean seeds. *Diss. Abst.* 38:1495.

Perl, M., Yaniv, Z., and Feder, Z. 1987. The effect of natural and accelerated aging on the lipid content and the fatty acid composition of seeds. *Acta Hort.* 215:61-67.

Roberts, E.H. 1972. *Viability of seeds.* Syracuse: Syracuse University Press, pp. 1-438.

Roberts, E.H. 1981. Physiology of aging and its application to drying and storage. *Seed Sci. Technol.* 9:359-372.

Roberts, E.H. 1983. Loss of seed viability during storage. *Advances in Research and Technology of Seeds* 8:9-34.

Roberts, E.H. and Abdalla, F.H. 1968. The influence of temperatures, moisture and oxygen on period of seed viability in barley, broad beans and peas. *Ann. Bot.* 32:97-117.

Sathiyamoorthy, P. and Nakamura, S. 1995. Free radical induced lipid peroxidation in seeds. *Israel J. Pl. Sci.* 43:295-302.

Sayre, J.D. 1940. Storage tests with seed corn. *Ohio J. Sci.* 40:181-185.

Schoettle, A.W. and Leopold, A.C. 1984. Solute leakage from artificially aged soybean seeds after imbibition. *Crop Sci.* 24:835-838.

Sokol, P.F. and Kononkov, P.F. 1975. The effect of high temperatures and air humidity on stored vegetable seeds. *Vestnik Sel'skokhozyaistvenoi Nauki Moscow* No. 7:16-22.

Stanwood, P.C. and Sowa, S. 1995. Evaluation of onion *(Allium cepa)* seeds after 10 years of storage at 5, – 18°C and – 196°C. *Crop Sci.* 35:852-856.

Steadman, K.J., Pritchard, H.W. and Dey, P.M. 1996. Tissue specific soluble sugars in seeds as indicators of storage category. *Ann. Bot.* 77:667-674.

Tao, K.L. 1992. Should vacuum packing be used for seed storage in gene banks? *Pl. Genetic Resources Newsletter* No.88/89:27-30.

Thanos, C.A. and Mitrakos, K. 1979. Phytochrome mediated germination control of maize caryopses. *Planta* 146:415-417.

Toole, E.H. and Toole, V.K. 1954. Relation of storage conditions to germination and to abnormal seedlings of beans. *Proc. Int. Seed Test. Assoc.* 18:123-129.

Trawartha, S.E., Tekrony, D.M., and Hildebrand, D.F. 1995. Soybean lipoxygenase mutants and seed longevity. *Crop Sci.* 35:862-868.

Vertucci, C.W. 1992. A calorimetric study of the changes in lipids during seed storage under dry conditions. *Pl. Physiol.* 99:310-316.

Villiers, T.A. and Edgcumbe, D.J. 1975. On the cause of seed deterioration in dry storage. *Seed Sci. Technol.* 3:761-764.

Yadav, T.D. and Mookherjee, P.B. 1975. Use of single, binary and tertiary gas mixture of nitrogen, CO_2 and oxygen in seed storage. *Seed Res.* 3:34-38.

Zhang, GungHua and Zhang, G.H. 1994. Ultra dry seed storage: Improved strategy and technology for germplasm conservation. *Chinese Biodiversity* 2:61-65.

Zhang, H.Y. and Kong, X.H. 1996. Tests on effect of vegetable seeds stored for many months. *Acta Agril. Bor. Sinica.* 11:118-123.

Zhang, M., Maeda, Y., Furihata, Y., Nakamaru, Y., and Esashi, Y. 1994. A mechanism of seed deterioration in relation to the volatile compounds evolved by dry seed themselves. *Seed Sci. Res.* 4:49-56.

SECTION II:
SEED STORAGE IN FRUIT CROPS

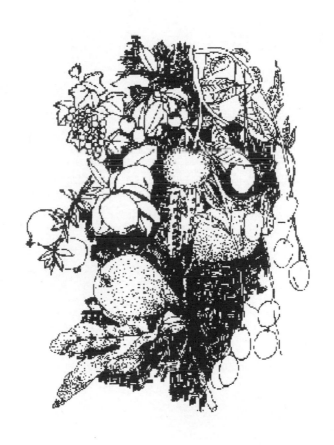

TROPICAL AND SUBTROPICAL FRUITS

– 2 –

Citrus Fruits (*Citrus* spp.)

Introduction

The major cultivated types of citrus fruits included in the genus *Citrus* are broadly grouped into oranges, mandarin, lemons, limes, grapefruit, and pummelo. The genus is cultivated for sweet and acidic fruits. Sweet orange, mandarin, and grapefruit are eaten fresh or processed for squash (sweetened citrus fruit juice) and juice preparation. Lemon and lime are acidic in nature and largely used in preparation of culinary products, such as pickles, and for flavoring food items. They are also processed for juice, squash, and lemonade. Citrus fruits are a rich source of sugar, citric acid, and vitamin C, and they possess valuable medicinal properties, being used in the prevention of colds and malaria, and to promote blood coagulation. The wide genetic diversity for citrus fruits exists in Southeast Asia, and they are predominantly distributed in the Indo-Chinese region. They grow well in tropical and subtropical regions. Some major citrus growing countries are Argentina, Australia, Brazil, China, Egypt, Greece, India, Israel, Italy, Japan, Mexico, Morocco, South Africa, Spain, Turkey, the United States, and the West Indies.

Morphology

Citrus fruits belong to family Rutaceae, subfamily Aurantioideae, and tribe Citreae. The tree is an evergreen, dicotyledonous perennial with spines. Leaves are simple ovate with wings at the petiole region. Flowers are single or in clusters, white, and fragrant. Fruit is a hesperidium, carpels, or segments filled with juicy arils and seeds. Seeds are white, show polyembryony, excepting *Citrus grandis,* and vary in size and number in different species.

Seed Storage

Citrus plants are propagated both sexually and asexually. Seeds of different *Citrus* species exhibit polyembryony, where nucellar seedlings are vigorous, true to type, and commonly used for raising citrus plants. In vegetative propagation, a large number of rootstocks is used to impart better fruit quality and resistance to various biotic and abiotic stresses in scionic plants. These are selected based on better compatibility with scion, hardy nature, and tolerance to saline conditions or soilborne diseases. They are raised exclusively through seeds. Seeds also play an important role in evolving new cultivars through hybridization, and in long-term conservation of genetic diversity.

High seed viability can be preserved for a shorter period for plant propagation, and for a longer period for the conservation of genetic diversity in the gene bank. Earlier, seeds of different *Citrus* species showed various storage behaviors, recalcitrance to orthodox under different storage conditions. However, King and Roberts (1979) grouped them as recalcitrant, wherein drying and freezing kill seeds. Drying of citrus seeds caused considerable delay in radicle emergence, thereby requiring a longer period for full seedling emergence. Further, rapid seed drying caused desiccation injury at higher temperatures (King and Roberts, 1980). Earlier seeds were germinated for a short period, and delayed seedling emergence due to seed drying was mistaken for loss of viability. In many *Citrus* species, seed germination is slow and erratic, owing to both physical and chemical inhibiting factors, and the undifferentiated embryo is likely to be a cause for delayed germination (Monselise, 1962). It is still unclear in certain *Citrus* species whether they belong to the orthodox or recalcitrant category. Hong and Ellis (1995) opined that interspecific variation for seed storage behavior exists within the genus *Citrus,* and the prevalence of an intermediate category of storage behavior is not ruled out between orthodox and recalcitrant types, similar to one that occurs in *Coffea* spp.

Seed Collections

Citrus is predominantly cultivated in subtropical regions, between 40° north and south of the equator. Young trees are susceptible to frost and grow well in sandy loam soil. Plants are raised almost exclusively by budding on desirable rootstocks and seldom through nucellar seedlings. The latter ones are more vigorous than sexual seedlings. *Citrus* species are cross-pollinated, and pollination is mainly by insects. Fruits develop slowly and take longer to mature. Seeds are extracted by cutting the fruits, then washed thoroughly, dried in shade or at low temperature, and suitably packed for storage. Washing seeds immediately after their extraction from fruits generally gives better germination than leaving seeds in rotting fruits. It was reported by Elze (1949) that the storage of picked fruits for several weeks affected germination.

SOUR ORANGE: *Citrus aurantium* L.

Sour orange, also called Seville orange, is commonly used in preparation of marmalade. Fruit is bitter and has medicinal properties. Sour orange is widely used as rootstock for sweet orange, lemon, and grapefruit due to its resistance to gummosis. It is a native of Southeast Asia and has somatic chromosome number 2n = 18. Tree is of medium size, growing to 10 meters (m) in height. Leaves are ovate, medium sized, and broadly winged. Flowers are large, axillary, white, and fragrant; stamens number 20 to 25; and the ovary has 10 to 12 locules. Fruit is subglobose, with a peel that is thick and rough, aromatic, and bright orange at maturity; the pulp is sour and bitter and embedded with many seeds. Seeds are polyembryonic and exhibit a high proportion of nucellar embryos.

Seed Storage

Seed Germination

Sour orange seeds germinate at a constant temperature ranging from 15 to 38°C (Camp, Mowry, and Loucks, 1933). Seedling emergence is rapid and higher at 30 to 35°C (Soetisna, King, and Roberts, 1985). According to Monselise (1953), optimum temperature for seed germination is 26°C. Seed viability is higher immediately after removal from ripe fruits. Seed shows progressive reduction in percentage of germination on air-drying or on storage (Spina, 1965). Burger (1983) noted that drying seed even for one day delays and decreases germination. Seed drying slowly at the rate of 30 percent moisture loss per 100 h imparted best viability (Edwards and Mumford, 1985a). Seeds selected from hardy trees are more vigorous and show higher percentage of polyembryony.

Seed germination is relatively poor in sour orange (Randhawa, Bajwa, and Bakshi, 1961) and is attributed to the presence of inhibiting substances (Monselise, 1959). Removal of testa and tegmen resulted in higher uptake of oxygen and gave higher percentage of germination (Demni and Bouzid, 1979; Mumford and Panggabean, 1982; Edwards and Mumford, 1985b). Furthermore, seed germination improves with seed soaking in water, in gibberellic acid at 50 parts per million (ppm) (Spina, 1965; Burger, 1983) and at 200 ppm (Yousif, Hassan, and Al-Sadoon, 1989), and in magnesium sulfate or magnesium citrate (1 percent) (Attalla and Haggag, 1987); with cold stratification (5°C for 45 days) (Rawash and Mougheith, 1978); and upon priming in polyethylene glycol (−0.90 megapascals [MPa] for 30 min) (Russo and Uggenti, 1994).

Storage Conditions

The viability of sour orange seeds decreases rapidly under ambient storage conditions (El-Wakeel and Ali, 1967), with seeds remaining viable for 120 days with or without seed coat removal (Mumford and Panggabean, 1982). Seed viability declines further on exposure of fruit to chilling temperatures (−7.8°C for 12 h) (Horanic and Gardener, 1959). It was earlier believed that seed loses viability rapidly on drying, and, therefore, storage in fruit juice, phenols, growth regulators, fungicides, and solutions of high osmotic potential would be also ineffective for preserving viability. However, seed stored at 4°C prevented germination and growth of microorganisms, and survival of imbibed seeds was best in 10^{-2} M (molar concentration) indoleacetic acid at 4°C (Edwards and Mumford, 1983). Edwards and Mumford (1985b) observed the longest survival of sour orange seeds in air at 4°C. Seed requires relatively small amount of oxygen to maintain viability during dry storage of seeds, and removal of moisture is beneficial in maintaining high viability for longer periods. King, Soetisna, and Roberts (1981) concluded that seed remains viable for eight months when dry seeds (5 percent mc) are stored at −20°C.

SWEET ORANGE: *Citrus sinensis* (L.) Osbeck

With its origin in Southern China sweet orange is a major fruit crop cultivated extensively in the citrus industry. The diploid chromosome number is 2n = 18. Fruit is eaten fresh as a dessert or snack. Tree is of medium size and grows up to 12 m high. Leaves are medium size with narrowly winged petioles. Flowers are axillary, single or in clusters, fragrant and white; stamens number 20 to 25; and the ovary has 10 to 14 locules. Fruit is subglobose and oval, with a tightly held peel, which turns orange on ripening, though it remains green in the tropics. Seeds are white, nil to many, and polyembryonic.

Seed Storage

Seed Germination

Sweet orange is propagated by vegetative means and requires rootstocks with straight roots and shoots for ideal grafting. Sweet orange seeds germinate at a temperature range of 15 to 37°C (Camp, Mowry, and Loucks, 1933). Seed germination is slow and often erratic. No germination of seeds occurs for the first 15 days of sowing, but subsequently it continues up to 31 days at 23°C (Kaufmann, 1969). Seed germination is better at an alternate temperature of 20/30°C for 16/8 h (Barton, 1943). Benching affects the uptake of nutrients and translocation of food reserve to fruits. The seed coat contains inhibitors, which was confirmed by inhibition of cress and lettuce seed germination (Panggabean and Mumford, 1982). Removal of the seed coat prevents benching of the stem (Brown et al., 1983), bending of roots (Said, 1967), and gives higher percentage of germination (Jan, 1956).

Storage Conditions

Higher seed moisture reduces the storage life. Well-dried seeds packed in polyethylene bags stored at 5.5°C maintain high viability (72 percent) for 12 months (Nauer and Carson, 1985). Seeds did not germinate when moisture content was less than 6 percent. The drier the seeds, the greater the delay in germination (Ferreira, 1969). Bajpai, Trivedi, and Prasad (1963) opined that leaving seeds in whole fruits is better than storage in polyethylene bags, and seed storage in charcoal is best for sweet orange. Seeds remained in good condition for several months when packed in sealed polyethylene bags (Anonymous, 1956). According to Horanic and Gardener (1959), low temperature affects seed viability during storage. Seeds stored at −3.9°C for 24 h were not injured but were partially damaged at −6.7°C and killed at −9.4°C.

MANDARIN: *Citrus reticulata* Blanco.

Mandarin is commonly referred to as loose-skinned orange. Fruit is sweet and subacidic and eaten fresh as a dessert or snack. It is also used for preparation of juice, squash, syrup, and lemonade. Mandarin originated in China, but its cultivation has extended to other tropical and subtropical regions. Mandarin is the hardiest among cultivated *Citrus* species. Trees are small to medium size, with a few thorns; leaves are small and have narrowly winged petioles. Flowers are small and white, stamens number 20, and the ovary has 10 to 15 locules. Fruit is depressed globose, with a thin peel; has easily separable segments; is orange-red, green, or yellow on ripening; and has sweet, juicy orange pulp. Seeds are small, pointed, and polyembryonic with green cotyledons. The somatic chromosome number is 2n = 18.

Seed Storage

Seed Germination

Mandarin seeds germinate at 15 to 35°C (Mobayen, 1980a), and a constant temperature of 25°C is ideal for germination (Monselise, 1953). Seed germination is affected by saline conditions, decreasing with increase in soil salinity (Caro et al., 1974). Seedling emergence is rapid and higher in seeds treated with ascorbic acid (100 ppm) (Misra and Verma, 1980).

Storage Conditions

Mandarin seeds survive for a shorter period under ambient conditions. High seed moisture at higher storage temperature promotes metabolism and encourages microbial growth in seeds, leading to loss of viability. Seeds of *Citrus* species behave differently toward desiccation and dry storage (Mumford and Panggabean, 1982). However, they are dried slowly and carefully as they attain a more advanced stage of germination; otherwise, they are killed on desiccation at a higher level of moisture (Farrant, Pammenter, and Berjak, 1988). Seed germination is reduced from 95 to 65 percent after two days of storage under room temperature, and seed moisture is concurrently reduced from 50.9 to 12.2 percent. Rapid reduction of germination might be due to desiccation injury (Doijode, unpublished data).

Loss of seed viability in *Citrus reticulata* is accompanied by a decline in nucleic acid levels; protein hydrolysis; protease, catalase, and peroxidase activity; and increases in leaching of metabolites (Shamsherry and Banerji, 1979). The quantum of leachate increases with an increase in storage period (see Figure 2.1).

Mungomery, Agnew, and Prodonoff (1966) suggested preserving mandarin seeds with initial moisture in a cool place and protecting them with suitable fungicides for longer storage. The percentage of germination was 24 in seeds stored at 4.5°C after 80 days of storage (Reddy, Sharma, and Singh, 1977). Seeds stored in fruit also retain the viability for 120 days at 15°C without affecting seed and fruit quality (Doijode, 1997) (see Figure 2.2).

LEMON: *Citrus limon* Burm.

Lemons are largely used in preparation of squashes, juices, lemonades, pickles, and various confectionery items. Lemon oil, citric acid, and pectin are extracted from the fruit and used in cosmetics. Lemon is native to Northeastern India. It is a small tree of about 3 to 6 m height with stiff thorns. The somatic chromosome number is 2n = 18. Leaves are simple and ovate and petioles are not winged. Flowers are single or in clusters, axillary, pink-

FIGURE 2.1. Changes in Viability and Vigor During Ambient Storage of Mandarin Seeds

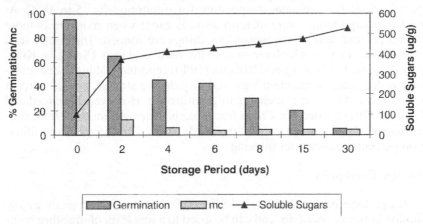

FIGURE 2.2. Seed Germination During Storage of Mandarin Fruits at Room Temperature and 15°C

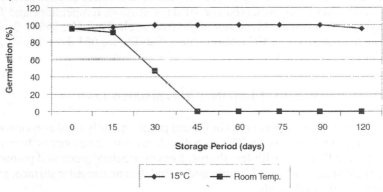

white; stamens number 20 to 40; and the ovary has eight to ten locules. Fruits are light yellow, oval, with a thick peel and terminal nipple, and sour. Seeds are ovoid and polyembryonic with white cotyledons.

Seed Storage

Seed Germination

Seed germination is slow and uneven in lemons due to their hard seed coat. The extract of lemon seed coat inhibits germination of cress and let-luce seeds (Panggabean and Mumford, 1982). Seed germination is further delayed and affected by drying, which requires longer imbibitions during germination. The optimum temperature for imbibitions is 35 to 40°C. At higher storage temperature, deterioration is rapid when moisture is more than 5 percent, and more abnormal seedlings are formed. The rate of seed germination increases with removal of the hard seed coat (Soetisna, King, and Roberts, 1985). King and Roberts (1980) reported that high seed moisture is necessary to maintain high viability during storage, and 22 percent mc or less reduces the percentage of germination. However, Mumford and Grout (1979) reported that *Citrus limon* seeds were tolerant to extreme desiccation (1.2 percent mc) and that dry seeds showed no changes in germination percentage even after freezing.

Storage Conditions

King, Soetisna, and Roberts (1981) confirmed that lemon seeds are orthodox in storage behavior and can be dried to a low level of moisture without affecting the viability. Furthermore, seeds with 5 percent moisture maintained viability for seven months at −20°C. Lemon seed viability is preserved well with lowering of seed moisture and under cool conditions. High seed viability is preserved for 40 days at room temperature and for 60 days at 5°C when seed is stored in the fruit itself (Doijode, 1984). Seeds without testa stored at −18°C retained their viability for 655 days (Mumford and Panggabean, 1982).

ROUGH LEMON: *Citrus jambhiri* Lush.

Rough lemon is indigenous to India and predominantly used as rootstock for cultivated *Citrus* species. The somatic chromosome number is 2n = 18. Trees are medium size with few thorns. Leaves are dark green and pointed; flowers are small; and fruit is rough and round, has an irregular surface, and contains many small seeds.

Seed Storage

Seed Germination

Seed viability is comparatively lower in rough lemon than in other rootstocks. Fresh seeds of Florida rough lemon cultivar gave only 20 percent germination (Bhekasut, Singh, and Sharma, 1976). Ali (1959) noted higher germination in descending order of rootstocks of rough lemon, smooth lemon, *Citrus karna, C. aurantifolia,* and *C. aurantium.* Seed germination can be improved by soaking seed in gibberellic acid (500 ppm)

(Singh, Shankar, and Makhija, 1979), and through cold stratification of seeds for 45 days at 5°C (Rawash and Mougheith, 1978).

Storage Conditions

Seed viability decreases with an increase in storage period. Seed germination reduced to 31 percent after four months of storage (Eshuys, 1975). Seed viability can be preserved for a longer period under cool and dry storage conditions. Seeds stored at 4°C and 96 percent RH exhibited 45 percent germination after 12 months of storage (Eshuys, 1974).

Invigoration of Stored Seeds

Seed vigor decreases during improper seed storage. Seed soaking in gibberellic acid (10 ppm) enhances germination from 31 to 76 percent even after four months of storage (Eshuys, 1975).

LIME: *Citrus aurantifolia* (Christm.) Swingle

Lime is widely cultivated in Asia for its edible fruits. Fruits are used for pickles, and their juice is used for flavoring food items and in the preparation of syrup and concentrate. Lime oil and citric acid are extracted from the fruit. Lime originated in northern parts of India and has somatic chromosome number 2n = 18. Lime trees are small with sharp spines, reaching up to 5 m in height. Leaves are small, ovate, with narrowly winged petioles. Flowers are small and white, stamens number 20 to 25, and the ovary has 9 to 12 locules. Fruits are small, round to oval, with a thin peel, aromatic, greenish yellow, and highly acidic. Seeds are white and polyembryonic.

Seed Storage

Seed Germination

Seed germination is higher at a constant temperature of 25°C (Panggabean, 1981). Large and extralarge fruits give bold seeds with higher germination and vigorous seedlings (Dharmalingam and Vijayakumar, 1987). The seed coat contains inhibitors that prevent germination, and its removal hastens the germination process (Panggabean and Mumford, 1982). Seed germination improves with soaking seeds in gibberellic acid (40 ppm), naphthalene acetic acid (40 ppm), thiourea (1.5 to 2.0 percent), and potassium nitrate (2 percent) (Choudhari and Chakrawar, 1980).

Storage Conditions

Seed longevity varies under different storage conditions. Mumford and Panggabean (1982) stored seeds for more than 120 days at room temperature. However, Doijode (1988) reported that germination declines rapidly on drying. It was 60 percent after 60 days of storage and lost completely by 365 days of ambient storage (see Figure 2.3).

High seed viability and vigor are preserved for 120 days when seeds are stored in fruit at 5 or 15°C (see Figure 2.4) (Doijode, 1998). Bajpai, Trivedi, and Prasad (1963) reported that lime seeds could be stored best in charcoal powder at 3°C and 50 percent RH. Lime seeds with 5 percent moisture stored at −20°C preserve viability for eight weeks (King, Soetisna, and Roberts, 1981). When storage is extended to 270 days at 4°C, viability is 90 percent (Sunbolon and Panggabean, 1986). Low moisture seeds (6.7 percent) stored in polyethylene and laminated aluminum foil pouches at 5 and −20°C preserve seed viability for three years. The retention of viability is greater in laminated aluminum foil pouches than in polyethylene bags (Doijode, 1990). Seed longevity can also be extended further with low-temperature (−20°C) storage, with 43 percent germination recorded after 12 years of storage (Doijode, unpublished data). Seed viability preserves well under modified atmosphere storage, especially in nitrogen (Doijode, 1996).

RANGPUR LIME: *Citrus limonia* Osbeck

Rangpur lime originated in India and is mainly used as a rootstock for citrus fruits in India and Brazil. The diploid chromosome number is 2n = 18. It is a hardy and drought-tolerant rootstock. Trees are medium sized with few thorns, with dull-green leaves and flowers that are tinged with purple. Fruits

FIGURE 2.3. Changes in Germination, Seed Moisture, and Dry Weight During Ambient Storage of Acid Lime Seeds

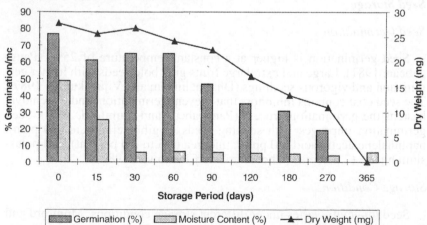

FIGURE 2.4. Seed Viability During Storage of Lime Fruits

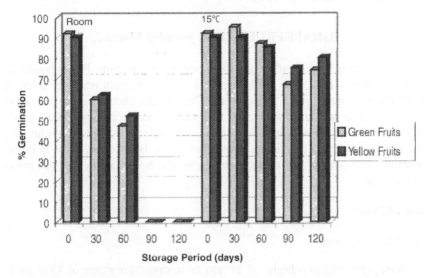

arc round to depressed globose with a short nipple, yellow to reddish orange, with a moderately loose peel; they have eight to ten segments, give
acidic juice, and contain numerous seeds. Large fruits contain a greater
number of heavy seeds, while unripe fruits contain few bold seeds (Desai
and Patil, 1983).

Seed Storage

Seed Germination

Seed germination is higher at a constant temperature range of 21 to 38°C
(Camp, Mowry, and Loucks, 1933). Seed germination decreases with fruit
size (Desai and Patil, 1983). Seeds are placed pointing downward or
sideways to get early and more straight seedlings (Moreira and Donadio,
1968). Seed viability and vigor reduce on drying. The degree of reduction
varies in different genotypes. It was higher in Brazilian rangpur lime, followed by the Florida strain. It appears that initial loss of viability is due to
desiccation injury (Doijode, 1994).

Storage Conditions

Rangpur lime seeds were stored better at 10°C and 45 percent RH for six
weeks (Kadam et al., 1994). Seed longevity can be extended further to

14 months when seeds are stored in sealed containers at 2 to 3°C and 33 to 38 percent RH (Bacchi, 1958).

GRAPEFRUIT: *Citrus paradisi* Macfad.

Grapefruit is largely cultivated for the edible juicy fruits. Fruits possess a characteristic flavor with mild bitterness and are largely used for juice. It probably originated in the West Indies. The somatic chromosome number is $2n = 18$. Trees are large and spiny with dense foliage of large, dark green leaves with winged petioles. Flowers are white, large, single or in clusters; stamens number 20 to 25; the ovary has 12 to 14 locules. Fruit is large, globose, and yellow and contains many seeds. Seeds are white and poly-embryonic.

Seed Storage

Seed Germination

Seed germination is higher at a constant temperature range of 15 to 38°C (Camp, Mowry, and Loucks, 1933) and at alternate temperatures of 20/30°C (16/8 h) (Barton, 1943). It improves with removal of seed coat (Tager and Cameron, 1957). Seeds extracted from immature fruit do not germinate (Fucik, 1974). The germination is slow and delayed on drying (Nauer, 1981), and it improves with soaking seeds in potassium dichromate at 0.5 percent (El-Wakeel and Ali, 1967).

Storage Conditions

Seed moisture plays an important role in deciding storage life of seeds. Low-moisture seeds stored in sealed conditions maintained viability for 16 weeks at room temperatures (Barton, 1943). Seeds stored in open conditions at room temperatures and at 5 and 8°C retained viability for three and four months, respectively. Seed viability over 50 percent can be retained for six months when seeds are stored at 5 and 8°C in sealed polyethylene and mixed with charcoal, ash, sand, or sawdust (Chacko and Singh, 1968). Seed germination was 40 percent when stored at 4.5°C in sealed plastic bags after 80 days of storage (Reddy, Sharma, and Singh, 1977). Seed viability was reduced on drying, whereas moist storage gave 80 percent germination (Nauer, 1981). Well-dried seeds packed in polyethylene bags stored at 5.5°C showed 82 percent germination after 28 months of storage (Nauer and Carson, 1985).

PUMMELO: *Citrus grandis* Osbeck

Pummelo is cultivated for edible fruits largely in East Asia. Fruits are eaten fresh and used in the preparation of mixed fruit drinks and marmalade. It is native to Malaysia and Thailand. Trees are medium in size and about 5 to 15 m in height. Leaves are large, ovate, and dark green. Flowers are large, cream colored, single or in clusters, and fragrant; stamens number 20 to 25; the ovary has 11 to 16 locules. Fruit is large, pear shaped or globose, and yellow; it has a thick peel and sweet juice; and it contains few seeds. Seeds are large, yellow, and monoembryonic. The diploid chromosome number is 2n = 18.

Seed Storage

Seed Germination

Pummelo seeds germinate well if they contain high moisture (15 to 20 percent). The reduction of seed moisture to 10 percent or less causes loss of viability (Nakagawa and Honjo, 1979). A minimum of 20 percent seed moisture is required to maintain high viability during storage of pummelo seeds (Honjo and Nakagawa, 1978).

Storage Conditions

Pummelo seeds need special attention during storage, being somewhat sensitive to drying. Also, a higher storage temperature reduces their viability. Seeds stored in sealed polyethylene bags at 4.4°C and 56 to 58 percent RH preserved their viability for 80 days (Reddy and Sharma, 1983).

TRIFOLIATE ORANGE: *Poncirus trifoliata* (L.) Raf.

Trifoliate orange is most commonly used as a hardy rootstock for citrus. It is cold hardy, induces dwarfing, and is fairly resistant to soilborne diseases and nematodes. It adapts to heavy soil and improves the scion fruit quality. Trifoliate orange fruits are nonedible, bitter, and used in medicine. Different plant parts are showy, and the plant is thus used in home gardens as an ornamental. It probably originated in Central and Northern China. It grows best in warm climates and in heavy soil. It does not grow well under saline conditions. Trees are small with thorns and three-lobed leaves; flowers are small, sessile, and white; stamens number 20 or more; the ovary has six to eight locules. Fruit is small, with a rough surface of lemon color and filled with acidic juice, and it contains numerous seeds. Seeds are ovoid, plump, and polyembryonic.

Seed Storage

Seed Germination

Trifoliate orange seeds germinate at various temperatures, ranging from 12 to 35°C. Seed germination is reduced by 20 to 25 percent at lower temperatures. Minimum and optimum temperatures are 6 to 10°C and 30°C, respectively (Mobayen, 1980a). Seed viability is affected by drying, and it decreases when moisture is reduced to less than 20 percent (Nakagawa and Honjo, 1979), 22 percent (Ferreira, 1969), and 27 percent (Montenegro and Salibe, 1960). The percentage of germination increases with removal of the seed coat (Okazaki, Ogawara, and Ino, 1964), washing in running water for 24 h (Singh and Soule, 1963), and soaking in gibberellic acid (10 ppm) (Eshuys, 1975). Seedling emergence is slow and delayed in fresh seeds, exhibiting 24 percent germination (Bhekasut, Singh, and Sharma, 1976).

Seed Maturity

Seed maturity coincides with fruit maturity. The percentage of germination is rather low before maturity. Trifoliate orange seeds collected 90 to 105 days after anthesis germinate. The germination is maximum at 135 days after flowering (Okazaki, Ogawara, and Ino, 1964). Embryo size plays a major role in seed germination. Reduction in embryo size lowers the percentage of germination (Singh and Soule, 1963). Germination of seeds extracted from yellow ripe fruits is comparatively lower than that for seeds of overripe fruits. However, seeds extracted at different stages of fruit development require storage for 12 weeks at 4.5°C for higher viability (Mobayen, 1980b).

Storage Conditions

Trifoliate orange seeds are short-lived, and their viability decreases with loss of moisture. A minimum of 25 percent moisture for storage of trifoliate seeds is recommended (Honjo and Nakagawa, 1978). Fresh seeds show 80 percent germination, and dry seeds stored at 4°C exhibited 3 percent germination, while moist seeds under similar conditions exhibited 50 percent germination after one year of storage (Chapot, 1955). Seeds with 27 percent moisture stored in plastic bags at 8°C retained viability for 224 days (Montenegro and Salibe, 1960). Seeds stored in a mixture of wet sand and vermiculite (50:50) at 5°C remained viable for 240 days but failed to germinate when stored in paper bags at ambient temperatures (Ferreira, 1970). Seeds stored at low temperature (4°C) and high humidity (96 percent RH) exhibited 32 percent germination after 12 months of storage (Eshuys, 1974). The phenolic substances increased in seeds during storage (Reddy and

Sharma, 1983). Okazaki, Ogawara, and Ino (1964) reported that seed longevity increased with open or sealed storage at low temperature or with storage in sand. Seeds stored in damp sand and at higher temperatures deteriorated rapidly. A well-dried seed material stored in polyethylene bags at 5.5°C gave 91 percent germination after 20 months of storage (Nauer and Carson, 1985).

REFERENCES

Ali, N. 1959. Studies on seed germination in rootstock species of *Citrus. Punjab Fruit J.* 22:22-32.

Anonymous. 1956. Transport of orange seeds from the Cook Island. *AR DSIR, NZ,* p. 16.

Attalla, A.M. and Haggag, M.N. 1987. Seed germination and growth and leaf composition of sour orange seedlings as influenced by pre-sowing treatments. *Alexandria J. Agril. Res.* 32:209-218.

Bacchi, O. 1958. Seed storage studies. II. Citrus. *Bragantia* 17:157-166.

Bajpai, P.N., Trivedi, R.K., and Prasad, A. 1963. Storage of citrus seeds. *Sci. Cult.* 29:45-46.

Barton, L.V. 1943. The storage of citrus seeds. *Boyce Thompson Inst. Contrib.* 13:47-55.

Bhekasut, C., Singh, R., and Sharma, B.B. 1976. Germination studies in citrus seeds. *Indian J. Hort.* 33:37-40.

Brown, S.C., Itae, C., Coombe, B.C., and Paleg, L.G. 1983. Germination of sweet orange seed. *Aust. Hort. Res. Newsletter* No. 55:104.

Burger, D.W. 1983. Germination and growth of seedlings from dried sour orange seeds. *J. Rio Grande Valley Hort. Soc.* 36:73-79.

Camp, A.F., Mowry, H., and Loucks, K.W. 1933. The effect of soil temperature on the germination of citrus seeds. *Amer. J. Bot.* 20:348-357.

Caro, M., Cerda, A., Fernandez, F.G., and Guillen, M.G. 1974. The tolerance of citrus rootstocks to salinity during germination. In *1st Congress Mundial de citricultura,* Ministerio de Agricultura Murcia, Spain 1:313-316.

Chacko, E.K. and Singh, R.N. 1968. Studies on the germination and longevity of fruit tree seeds—*Citrus* spp. *Indian J. Hort.* 25:94-103.

Chapot, H. 1955. Remarks on the germination of seeds of *Poncirus trifoliate. Fruits d'Outre Mer.* 10:465-468.

Choudhari, B.K. and Chakrawar, V.R. 1980. Effect of some chemicals on the germination of Kagzi lime *(Citrus aurantifolia)* seeds. *J. Mah. Agril. University* 5:173-174.

Demni, S. and Bouzid, S. 1979. First results on the germination of sour orange seeds. *Fruits* 34:283-287.

Desai, J.S. and Patil, V.K. 1983. Influence of fruit size on seed germination in Rangpur lime *(Citrus limonia). Seed Res.* 11:68-70.

Dharmalingam, C. and Vijayakumar, A. 1987. Seed quality in relation to fruit size in acid lime *(Citrus aurantifolia). South Indian Hort.* 35:274-279.

Doijode, S.D. 1984. Prolonged viability of lemon seeds preserved in fruit. *South Indian Hort.* 32:151-152.

Doijode, S.D. 1988. Longevity of Kagzi lime seeds under ambient storage conditions. *Punjab Hort. J.* 28:114-117.

Doijode, S.D. 1990. Seed viability and longevity under different storage temperatures and containers in acid lime *(Citrus aurantifolia). Inter. Conf. Seed Sci. Technol. New Delhi,* Abstr. No. 2.46, p. 55.

Doijode, S.D. 1994. Changes in seed germinability on drying in certain strains of Rangpur lime. *XXIV Inter. Hort. Congr.* Abstr. 82:31.

Doijode, S.D. 1996. Preservation of Kagzi lime seeds in controlled atmosphere. *Pl. Physiol. Biochem.* 23:96-98.

Doijode, S.D. 1997. Seed quality and seed storage in fruit for short-term conservation of germplasm in Coorg mandarin *(Citrus reticulata). Plant Genetic Resources Newsletter* 111:82-83.

Doijode, S.D. 1998. Seed viability in relation to storage of fruits under different temperatures in acid lime. *Haryana J. Hort. Sci.* 27:5-7.

Edwards, C.A. and Mumford, P.M. 1983. Storage of *Citrus aurantium* seeds imbibed in chemicals. *Plant Genetic Resources Newsletter* No. 56:13-22.

Edwards, C.A. and Mumford, P.M. 1985a. The effect of the drying methods on the behavior of sour orange *(Citrus aurantium)* seeds. *Seed Sci. Technol.* 13:227-234.

Edwards, C.A. and Mumford, P.M. 1985b. Factors affecting the oxygen consumption of sour orange *(Citrus aurantium)* seeds during imbibed storage and germination. *Seed Sci. Technol.* 13:201-212.

El-Wakeel, A.T. and Ali, N.A. 1967. Soaking of citrus seeds. *Agric. Res. Rev. Cairo* 45:81-85.

Elze, D.L. 1949. Germination of citrus seeds in relation to certain nursery practices. *Palest. J. Bot.* 7:69-80.

Eshuys, W.A. 1974. Loss of viability of *Citrus* and *Poncirus trifoliata* seeds during storage. *Citrus and Sub-Trop. Fruit J.* No. 489:5-7.

Eshuys, W.A. 1975. The effect of GA on the germination of citrus seeds. *Information Bull. Citrus and Sub-Trop. Fruit Res. Inst.* No. 32:3-4.

Farrant, J.M., Pammenter, N.W., and Berjak, P. 1988. Recalcitrance—A current assessment. *Seed Sci. Technol.* 16:155-166.

Ferreira, J.J. 1969. Loss of germination capacity in *Citrus* spp. *Rev. Fac. Agron. Vet. B. Aires* 17:51-55.

Ferreira, J.J. 1970. The influence of distinctive storage conditions on trifoliate orange seed germination. *Rev. Fac. Agron. Vet. B. Aires* 18:59-63.

Fucik, J.E. 1974. Grapefruit seed viability and germination. *J. Rio Grande Valley Hort. Soc.* 28:140-142.

Hong, T.D. and Ellis, R.H. 1995. Interspecific variation in seed storage behavior within two genera—*Coffea* and *Citrus. Seed Sci. Technol.* 23:165-181.

Honjo, H. and Nakagawa, Y. 1978. Suitable temperature and seed moisture content for maintaining the germinability of citrus seeds for long-term storage. In long-term preservation of favorable germplasm in arboreal crops. *Ibaraki ken Japan, Fruit Tree Res. Stn. Ministry Agril. Forestry,* pp. 31-35.

Horanic, G.E. and Gardener, F.E. 1959. Effects of sub-freezing temperature on the viability of citrus seeds. *Citrus Industry* 40:12A-12B.

Jan, S. 1956. The effect of removing the seed coat and of chemical treatment of seeds of *Citrus sinensis* var. Sekkan and *Euphoria longana* on germination and early stages of seedling development. *Acta Agric. Sinica* 7:465-473.

Kadam, A.S., Khedkar, D.H., Patel, V.K., and Anserwadekar, K.W. 1994. Studies on viability and germinability of Rangpur lime seeds during storage. *J. Mah. Agril. Universities* 19:130-131.

Kaufmann, M.R. 1969. Effect of water potential on germination of lettuce, sunflower and citrus seeds. *Canadian J. Bot.* 47:1761-1764.

King, M.W. and Roberts, E.H. 1979. *The storage of recalcitrant seeds—Achievements and possible approaches.* International Board for Plant Genetic Resources, pp. 1-96.

King, M.W. and Roberts, E.H. 1980. The desiccation response of seeds of *Citrus limon. Ann. Bot.* 45:489-492.

King, M.W., Soetisna, U., and Roberts, E.H. 1981. The dry storage of citrus seeds. *Ann. Bot.* 48:865-872.

Misra, R.S. and Verma, V.K. 1980. Studies on the seed germination of Kinnow orange in the Central Himalayas. *Prog. Hort.* 12:79-84.

Mobayen, R.G. 1980a. Germination and emergence of citrus and tomato seeds in relation to temperature. *J. Hort. Sci.* 55:291-297.

Mobayen, R.G. 1980b. Germination of trifoliate orange seed in relation to fruit development, storage, and drying. *J. Hort. Sci.* 55:285-289.

Monselise, S.P. 1953. Viability tests with citrus seeds. *Palestine J. Bot.* 8:152-157.

Monselise, S.P. 1959. *Citrus* germination and emergence as influenced by temperature and seed treatments. *Bull. Res. Coun. Israel Sect. D.* 7D:29-34.

Monselise, S.P. 1962. Citrus seed biology. *XVIth Inter. Hort. Congr. Brussels,* pp. 559-565.

Montenegro, H.W.S. and Salibe, A.A. 1960. Storage of seeds of citrus rootstocks. *Rev. Agric. Piracicaba* 35:109-135.

Moreira, C.S. and Donadio, L.C. 1968. The effect of the position of the planted seeds on the type of seedling produced. *Solo* 30:69-70.

Mumford, P.M. and Grout, B.W.W. 1979. Desiccation and low-temperature (−196°C) tolerance of *Citrus limon* seeds. *Seed Sci. Technol.* 7:407-410.

Mumford, P.M. and Panggabean, G. 1982. A comparison of the effects of dry storage on seeds of *Citrus* species. *Seed Sci. Technol.* 10:257-266.

Mungomery, U.V., Agnew, G.W.J., and Prodonoff, E.T. 1966. Maintenance of citrus seed viability. *Qd. J. Agric. Anim. Sci.* 23:103-120.

Nakagawa, Y. and Honjo, H. 1979. Studies on the freezing hardiness and drought tolerance of citrus seeds. *Bull. Fruit Tree Res. Stn.* No. 6:27-35.

Nauer, E.M. 1981. Drying of rootstock seeds delay germination. *Citrograph* 66:204-206.

Nauer, E.M. and Carson, T.L. 1985. Packaging citrus seeds for long term storage. *Citrograph* 70:229-230.

Okazaki, M., Ogawara, K. and Ino, Y. 1964. The germinability and storage of *Poncirus trifoliata* seeds. *Sci. Rep. Fac. Agric. Okayama* No. 24:29-36.

Panggabean, G. 1981. Dry storage of citrus seeds and factors affecting their germination. MSc. thesis. University of Birmingham, Birmingham, AL.

Panggabean, G. and Mumford, P.M. 1982. The inhibitory effects of seed coats on citrus seed germination. *Ann. Bogorienses* 7:167-176.

Randhawa, S.S., Bajwa, B.S., and Bakshi, J.C. 1961. Studies on the behavior of seedlings of some common citrus rootstocks in nursery. *Indian J. Hort.* 18:71-80.

Rawash, M.A. and Mougheith, M.G. 1978. Effect of some storage treatments on seed germination of some citrus rootstocks. *Res. Bull. Ain Shams University Fac. Agr.* No. 835:11.

Reddy, M.K., Sharma, B.B. and Singh R. 1977. Changes in germinability of citrus seeds during storage. *Seed Res.* 5:145-151.

Reddy, M.K. and Sharma, B.B. 1983. Effect of storage conditions on germination, moisture content and some biochemical substances in citrus seeds. II. Trifoliate orange and pummelo. *Seed Res.* 11:56-59.

Russo, G. and Uggenti, P. 1994. Osmotic priming in ecotypes of *Citrus aurantium* L. seeds to increase germination rate, seed polyembryony and seedling uniformity. *Acta Hort.* 362:235-241.

Said, M. 1967. Effect of captan on benching of roots and germination period of citrus seedlings. *Agric. Pakistan* 18:453-456.

Shamsherry, R. and Banerji, D. 1979. Some biochemical changes accompanying loss of seed viability. *Pl. Biochem. J.* 6:54-63.

Singh, B.P. and Soule, M.J., Jr. 1963. Studies in polyembryony and seed germination of trifoliate orange *(Poncirus trifoliata). Indian J. Hort.* 20:21-29.

Singh, H.K., Shankar, G., and Makhija, M. 1979. A study on citrus seed germination as affected by some chemicals. *Haryana J. Hort. Sci.* 8:194-195.

Soetisna, U., King, M.W., and Roberts, E.H. 1985. Germination tests recommendation for estimating the viability of moist or dry seeds of lemon *(Citrus limon)* and lime *(Citrus aurantifolia). Seed Sci. Technol.* 13:87-110.

Spina, P. 1965. Studies on the germinating capacity of sour orange seeds. *Atti. Giorn Stud. Prop. Spec. legn. Pisa,* pp. 407-417.

Sunbolon, H. and Panggabean, G. 1986. Some aspects of citrus with special reference to Indonesia. *Bull. Pene. Hort.* 14:32-40.

Tager, J.M. and Cameron, S.H. 1957. The role of the seed coat in chlorophyll deficiency of citrus seedlings. *Physiol. Plant.* 10:302-305.

Yousif, Y.H., Hassan, K., and Al-Sadoon, H.S. 1989. Effect of gibberellic acids on germination of sour orange seeds and their growth in ten soil mixes. *Ann. Agril. Sci.* 34:1139-1149.

3

Mango: *Mangifera indica* L.

Introduction

Mango is an important tropical fruit largely cultivated in Southeast Asia for its delicious fruit. It is most popular in India. Ripe fruits are eaten fresh and used in various products, such as juice, squash, jams, and jellies, and products such as chutney, pickles, and curries are made from unripe fruits. Fruits are rich in vitamins A and C, and kernels contain high amounts of carbohydrates, calcium, and fat.

Origin and Distribution

Mango originated in the Indo-Burma region. *Mangifera* species are distributed throughout the Indo-Malayan region, showing wide genetic diversity for different plant characters (Mukherjee, 1972). Subsequently it was introduced to other tropical countries. It is cultivated in Australia, Bangladesh, Brazil, China, Colombia, Cuba, Egypt, Fiji, India, Indonesia, Jamaica, Madagascar, Malawi, Malaysia, Mauritius, Myanmar, Nigeria, Peru, the Philippines, Portugal, Seychelles, South Africa, Sri Lanka, Sudan, Thailand, the United States, Venezuela, and the West Indies.

Morphology

Mango belongs to family Anacardiaceae and has chromosome number 2n = 40. It is a medium-sized, evergreen, dicotyledonous, perennial tree. Leaves are simple and dark green. Inflorescence is large and terminal; panicle bears about 300 to 1,000 staminate and perfect flowers. Fruit is an indehiscent, fleshy drupe and varies in size, shape, and color, being either green, yellow, or red. Pulp is soft, flavored, juicy, and green, yellow, or orange in color. Seeds are enclosed in a hard endocarp, with two fleshy cotyledons and little or no endosperm. Most of the cultivars are monoembryonic, while a few show polyembryony, with 2 to 12 embryos, where both apomictic and zygotic embryos exist.

Seed Storage

Mango is commercially propagated by asexual methods. However, seeds are used for raising uniform rootstocks and orchards with polyembryonic cultivars, ensuring true-to-type plants. Seedling trees are long-lived and deep-rooted and show large variations in bearing habit and fruit characteristics. They are relatively late bearers and difficult to manage in large numbers. Normally, seeds are used for raising rootstocks and in hybridization programs. Mango seeds remain viable for short periods and exhibit recalcitrant storage behavior. Seeds are susceptible to desiccation and are killed when moisture goes below a critical level. Further, they do not withstand chilling temperature during storage. Seeds stored under moist conditions remain viable only for a short period, from a few weeks to a few months.

Seed Collections

Mango grows well in tropical conditions. Young and flowering trees are damaged by frost. It can grow up to 1,000 m altitude in areas but performs best up to 500 m altitude. It thrives well in deep sandy loam soils. It is commonly propagated asexually through graftage. In polyembryonic cultivars, seeds are used for raising plants as well as rootstocks. Mango is cross-pollinated by insects. Fruits are collected from healthy trees having the desired qualities (see Figure 3.1). Fully ripe and healthy fruits are selected, and the pulp is removed from the stones, washed thoroughly, and stored.

Seed Germination

Mango seeds are nondormant and germinate readily under favorable conditions. They are enclosed by a hard endocarp, which inhibits or delays seedling emergence. Removal of seeds from stones increases germination (Chauran et al., 1979). According to Sinnadurai (1975), decorticating helps in early emergence of seedlings and is also beneficial in development of the nucellar seedlings in polyembryonic cultivars. Seeds of monoembryonic cultivars germinate readily and emerge earlier than those of polyembryonic cultivars (Bakshi, 1963). Seeds extracted from soft-pulp fruits give higher germination (76 percent) than those from firm fruits (54 percent), and seeds from firm fruits remain dormant for some time after planting (Giri, 1966). Corbineau, Kante, and Come (1987) reported that germination occurs between 5 and 40°C, but the optimum temperature is 25 to 30°C. Seedlings grow well at 30°C, and higher temperature (40°C) is injurious to growing seedlings, while temperature below 15°C causes chilling injury. Giri and Chaudhary (1966) noted that seed germination and seedling vigor are posi-

FIGURE 3.1. Mango Tree with Many Fruits

tively related to seed weight. Large embryos give rapid emergence of vigorous seedlings (Corbineau, Kante, and Come, 1987).

Storage Conditions

Mango seeds lose viability if dried immediately after extraction. Due to their recalcitrant nature, presently they can only be preserved for a short duration and under specific conditions. A critical level of moisture determines the longevity of seeds, and this varies in cultivars, stages of maturity, and growing conditions. Doijode (1990) reported that reduction of seed moisture below 25 percent in 'Alphonso' and 32 percent in 'Totapuri' lowered the percentage of germination and seedling vigor (see Figure 3.2). Fresh seeds remain viable for four to five weeks under ambient conditions (Singh, 1960; Doijode, 1990), while Ito and Atubra (1973) reported that seeds remained viable for ten days under dry storage. Loss of viability is greater in high-moisture seeds, and also at higher storage temperatures. Parisot (1989) opined that loss of water from stored seeds and loss of germination capacity were rapid at high storage temperature (20 to 30°C). Retention of high moisture during storage promotes fungal growth, which causes discoloration and loss of viability. High moisture also stimulates the sprouting of seeds during ambient storage. With these constraints in view, suitable storage conditions need to be created, but so far, not much has been achieved in this regard.

Seed storage in fruit. Mango seeds deteriorate rapidly on separation from the fruit. As long as seeds remain inside the fruit, they are germinable. How-

FIGURE 3.2. Changes in Seed Moisture and Viability During Seed Storage in Mango

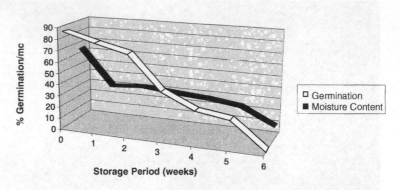

ever, subsequent shriveling and rotting of fruits affect the seed quality during ambient storage. Storing fruits at low temperature causes chilling injury. According to Simao (1959), the best method is to store whole fruit at room temperature (22 to 25°C), allowing seeds to remain viable for 70 days. Storage at 5°C harms the fruit. Large-fruit seeds stored better than small-fruit seeds.

Seed storage with charcoal. Charcoal powder is a better storage medium for mango seeds than polyethylene bags, desiccators, refrigeration, or open jars. It maintains suitable moisture and protects the seeds from pathogens. With this, viability of 50 to 60 percent is retained up to 90 days (Teaotia and Singh, 1971). Bajpai and Trivedi (1961) reported that seeds stored with charcoal in a desiccator at 50 percent RH gave 80 percent germination even after 70 days of storage. Likewise, Chacko and Singh (1971) obtained 37.5 percent germination after 100 days of storage of seeds with charcoal at 20 to 25°C. However, viability was reduced to 17.5 percent after 120 days of storage. Similarly, about 40 percent of seeds germinated after 90 days of storage when stones were stored in polyethylene bags along with charcoal at 25°C. While in open conditions, only 12 percent of seeds germinated by the sixtieth day of storage (Patil, Gunjate, and Salvi, 1986).

Seed storage in moist conditions. Seeds stored under moist conditions at relatively lower temperatures preserve viability. Seeds of cultivar Dashehari, when stored in thin polyethylene bags (0.025 millimeters [mm]) after treatment with 1 percent 8-hydroxyquinolene sulfate at 15°C, showed 50 percent germination after 12 months of storage (see Figure 3.3) (Doijode, 1995). Seeds stored at 8°C did not germinate (Patil, Gunjate, and Salvi, 1986), and chilling injury occurs at very low temperatures (Corbineau and

FIGURE 3.3. Moist Storage of Mango Seeds

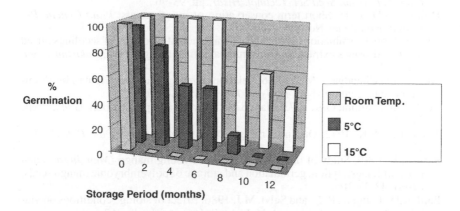

Come, 1988). The loss of seed viability is associated with structural changes in cells during deterioration, causing greater leakage of soluble sugars and amino acids and a decrease in dehydrogenase activity (Doijode, 1989). Furthermore, Chandra (1980) reported a decline in RNA and reducing sugar in dry seeds with aging. Mango seeds having moisture above 20 percent maintain high viability at 15°C when packed in thin polyethylene bags and suitably protected against storage fungi.

REFERENCES

Bajpai, P.N. and Trivedi, R.K. 1961. Storage of mango seed stones. *Hort. Adv.* 5:228-229.

Bakshi, J.C. 1963. Germination of mango stones in relation to the depth and time of sowing. *Punjab Hort. J.* 3:199-204.

Chacko, E.K. and Singh, R.N. 1971. Studies on the longevity of papaya, phalsa and mango seeds. *Proc. Int. Seed. Test. Assoc.* 36:147-158.

Chandra, N., 1980. Some physiological changes accompanying loss of viability of the seeds of *Mangifera indica. Plant Biochem. J.* 7:105-109.

Chauran, O.R., Manica, I., Pinheiro, R.V.R., Conde, A.R. and Chaves, J.R.P. 1979. The effect of storage time, seed treatment, and fungicides on seed germination and growth of mango seedlings. *Revista Ceres.* 26:1-12.

Corbineau, F. and Come, D. 1988. Storage of recalcitrant seeds of four tropical species. *Seed Sci. Technol.* 16:97-103.

Corbineau, F., Kante, M., and Come, D. 1987. Seed germination and seedling development in mango. *Fruits* 42:113-120.

Doijode, S.D. 1989. Loss of viability and membrane damage during storage of mango seeds. *Proc. Natl. Sem. Mango Lucknow.* 1:24.

Doijode, S.D. 1990. Influence of seed moisture on longevity of mango seeds. *Proc. Inter. Sat. Symp. Seed Sci. Technol. Hisar,* pp. 93-96.

Doijode, S.D. 1995. Short term conservation of mango seeds. *Plant Genetic Resources Newsletter* No. 104:24-25.

Giri, A. 1966. Germination percentage, average height, and girth of seedlings raised from seed stones extracted from syrupy and firm mango fruits. *Pakistan J. Sci.* 18:79-81.

Giri, A. and Chaudhary, M.Y. 1966. Relation of mango stone's weight to its germination and seedling vigor. *Pakistan J. Sci.* 18:148-150.

Ito, P.J. and Atubra, O.K. 1973. Mango seed storage and propagation. *Ghana Fmr.* 17:34-38.

Mukherjee, S.K. 1972. Origin of mango *(Mangifera indica). Econ. Bot.* 26:260-264.

Parisot, E. 1989. Study of the growth rhythm in young mango *(Mangifera indica* L.). Part I. Description, germination and storage of polyembryonic mango seeds. *Fruits* 43:97-105.

Patil, R.D., Gunjate, R.T., and Salvi, M.J. 1986. Effect of storage conditions on viability of mango seed stones. *J. Mah. Agril. Universities* 11:362.

Simao, S. 1959. A study of the germinating power of mango seeds. *An. Esc. Sup. Agric. Queiroz.* 16:289-297.

Singh, L.B. 1960. *The mango—Botany, cultivation and utilization.* London: Leonard Hill, Book Company.

Sinnadurai, S. 1975. The effect of decortications of seeds on germination and the number of nucellar seedlings of some mango cultivars in Ghana. *Acta Hort.* 49:95-97.

Teaotia, S.S. and Singh, R.D. 1971. Studies on media for storage and germination of mango seed stones *(Mangifera indica). Punjab Hort. J.* 11:52-56.

4

Papaya: *Carica papaya* L.

Introduction

Papaya, also called pawpaw or melon tree, is a major dessert fruit crop in the tropics. Rich in carbohydrates, carotene, riboflavin, and vitamin C content, the fruit's sweet-flavored pulp is eaten fresh. It is also processed for jam, jelly, candy, and juice preparations. Different plant parts, particularly of the unripe fruit, yield a milky latex that contain the proteolytic enzyme papain, which has many industrial and medicinal uses. Unripe fruits are also used for pickle preparation and cooked as vegetables.

Origin and Distribution

Papaya originated in Central America. Many of its wild species are grown in Colombia, Ecuador, Mexico, and Nicaragua. The papaya is cultivated throughout the tropical and subtropical regions. Some major papaya-growing countries are Australia, Brazil, Colombia, Costa Rica, Cuba, Ecuador, India, Kenya, Malaysia, Mexico, Myanmar, Nigeria, Peru, the Philippines, Seychelles, South Africa, Sri Lanka, Taiwan, the United States, and Venezuela.

Morphology

Papaya belongs to family Caricaceae; its somatic chromosome number is 2n = 18. Papaya is a small herbaceous and dicotyledonous plant with a stem that is soft wooded, hollow, and straight; rarely branches; and ends with a cluster of foliage. Leaves are large lobed with long petioles. Plants show a pistillate habit that is stable, while staminate and andromonoecious trees produce staminate, pistillate, and hermaphrodite flowers, depending on climatic conditions. Storey (1969) reported 31 sex types in papaya, with 15 each for variation in staminate and andromonoecious trees, and the final one pistillate. Staminate and andromonoecious trees are heterozygote, while the pistillate is a recessive homozygote. Fruit is a berry that is ovoid

47

and oblong to spherical or cylindrical in shape and can be or grooved. Flesh is yellow to reddish orange. Fruits vary in size and shape depending upon sex type and cultivars. Fruit cavity is filled with a mass of black, round seeds about 5 mm in diameter, surrounded by gelatinous sarcotesta; the embryo is straight with ovoid flattened cotyledons enclosed by fleshy endosperm. Seeds constitute about 16 percent of total fresh weight of fruit (Passera and Spettoli, 1981).

Seed Storage

In papaya, seeds are predominantly used for raising plants. They are also used for evolving new cultivars through hybridization, and for long-term conservation of genetic diversity. Papaya seeds survive for a short period under ambient conditions. Seeds show orthodox storage behavior (Ellis, 1984), in which seeds withstand desiccation and extend longevity at lower temperatures. Papaya seeds, however, are susceptible to chilling temperatures and killed when stored at zero or subzero temperatures. Therefore, seeds may be considered intermediate to recalcitrant with orthodox storage behavior.

Seed Collections

Papaya is a warm-season crop. It requires hot sunny days for production of high-quality fruits. It is susceptible to frost, and low temperatures affect growth. Papaya grows well in deep, fertile, well-drained clay loamy soil that is free from nematodes (Salazar-Castro, 1982). Either seeds are sown directly in the field or seedlings are raised in a nursery and subsequently transplanted in field. Seedling emergence is faster and higher when the plot is covered with cheesecloth (Camejo and Rivero, 1987). Flowering commences four to eight months after planting, and insects cause pollination. To maintain genetic purity of strains, flowers are hand pollinated and covered with bags to prevent pollen contamination. Fruit takes about five months to mature. It turns green to yellow on maturity (see Figure 4.1). A greater fruit weight produces more seeds in fruits ripened during spring and autumn than in those ripened during summer (Allan, 1969). The seed mass is soaked in water for 24 h, and later seeds are separated from the pulp. Seeds are cleaned; mucilaginous material is removed, dried, and suitably packed for storage.

Fruit maturity. Seed germination and longevity are dependent on seed maturity, which coincides with maturity of the fruit. Seed matures when the fruit becomes yellowish. Seed germination was highest (80 percent) in seeds collected from the least mature fruits (one-quarter mature) and stored for 33 days at room temperature (Limadiaz et al., 1985).

FIGURE 4.1. Papaya Fruits Ready for Harvest

Seed drying. Rapid drying of seeds affects germination, whereas sun drying does not affect seed viability (Chacko and Singh, 1971). Longer duration of sun drying (>3 days) reduces germination. Seedling emergence is earlier and higher in seeds dried in shade compared to those dried in sun (Vecchio and Shirwa, 1987).

Seed Germination

Fresh papaya seeds do not germinate readily. They exhibit dormancy up to a period of 35 days (Palaniswamy and Ramamoorthy, 1987). In fresh seeds, seedling emergence is slow and nonuniform (Chacko and Singh, 1966) and only 6 percent of seeds germinate (Koyamu, 1951). Low seed germination is also partly due to the absence of an embryo in about 20 percent of seeds (Nagao and Furutani, 1986), and seed viability improves with a short period of storage. Large seeds give a slightly higher germination percentage than small seeds (Singh and Singh, 1979). Papaya seed germination is influenced by the following factors.

Sarcotesta. This is a gelatinous mucilaginous substance derived from the aril that surrounds papaya seeds. It protects the seeds from extraneous environment and delays germination. Seed germination is slower and lower in seeds with the sarcotesta intact than in seeds from which it is removed (Lange, 1961b; Saito and Yamamota, 1965; Mosqueda, 1969; Perez, Reyes, and Cuevas, 1980; Reyes, Perez, and Cuevas, 1980). Fresh seeds with the aril intact did not germinate for 30 days after sowing. Removal of the aril from seeds enhances germination to 40 to 50 percent, and further air-drying increases germination to above 80 percent (Yahiro, 1979). Reyes, Perez, and Cuevas (1980) reported the presence of inhibitors in sarcotesta and sclerotesta that affected germination. Seeds soaked in seed leachate or extracts of sarcotesta or in papaya fruit juice germinate poorly, and this shows the presence of inhibitors (Saito and Yamamota, 1965). Simple washing of seeds removes inhibitors to a certain extent and improves seed germination (Mosqueda, 1969; Perez, Reyes, and Cuevas, 1980).

Temperature. Papaya seed germination is influenced by temperature, with an optimum temperature of around 35°C. Temperatures below 23°C and above 44°C are detrimental to germination. Night temperatures are more critical for seedling growth than day temperatures or day length (Lange, 1961a).

Chemical and growth substances. Endogenous growth promoters are found in the embryo and endosperm, while inhibitors are present in the sarcotesta. Seedling emergence improves with removal of the sarcotesta and also by soaking seeds in potassium nitrate (1.0 M), gibberellic acid (600 ppm), and thiourea (Nagao and Furutani, 1986). Higher concentrations of thiourea (4 to 10 grams per liter [g·liter^{-1}]) affected germination and vigor (Begum, Lavania, and Babu, 1988). Seeds treated with gibberellic acid gave rapid emergence of seedlings (Furutani and Nagao, 1987) and higher germination, particularly in seeds without sarcotesta (Saito and Yamamota, 1965).

Soil salinity. Seed germination is delayed and seedling growth is adversely affected by the presence of higher salt concentrations (Hewitt, 1963). Seedlings are slender, stunted, and cannot survive long under saline conditions beyond 4 millimhos per centimeter (mmhos/cm) (Makhija and Jindal, 1983).

Storage Conditions

Papaya seeds deteriorate rapidly at higher storage temperatures and relative humidity. They exhibit intermediate to recalcitrant and orthodox storage behavior. Fresh seeds give higher germination and seedling vigor that will decline on storage. The seed longevity is greatly influenced by type of container and storage conditions (Storey, 1969; Bass, 1975; Kalie and Hartiningsih, 1973).

Temperature. Papaya seeds maintain viability for three months under ambient temperatures (Begum, Lavania, and Babu, 1988). Seed viability decreases at lower storage temperatures. Seeds stored at room temperature give higher germination than seeds stored in a desiccator or refrigerator (Saito and Yamamota, 1965). Singh and Singh (1981) reported that seeds remained viable for eight months at room temperature when packed in sealed polyethylene bags. Seed germination decreases with increase in storage period. Higher seed viability and vigor are maintained at 4 to 6°C than at room temperatures. The loss of viability was attributed to accumulation of inhibitors in seeds (Reyes, Perez, and Cuevas, 1980). Seeds stored at 10°C and 50 percent RH in cloth bags or at 5°C in sealed moisture-proof packages preserved viability reasonably well during six years of storage (Bass, 1975). High seed germination is recorded in seeds stored at 10°C after two years of storage. However, seed germination is affected when stored at 5°C and −18°C (see Figure 4.2) (Doijode, 1993).

Seed packaging. Packaging is very important to protect the seeds from higher humidity and insects or pathogens. The selection of packaging depends on duration and conditions of storage. Loss of viability is greater in open storage than in sealed storage using polyethylene bags or glass jars (Chacko and Singh, 1971). Seed viability can be preserved for 11 months when seeds are mixed with ash and stored in a capped blue bottle (Kalie and Hartiningsih, 1973). Seed germination is higher (57 percent) in seeds stored in plastic bags at 4 to 6°C after 29 months of storage, and seeds stored in cloth bags at room temperature gave the lowest germination (16 percent) (Singh and Singh, 1981). Seeds stored with nitrogen gas in laminated aluminum foil pouches exhibited higher germination during ambient storage (Doijode, 1992). Seed germination was higher in seeds stored in paper bags, followed by seeds stored in polyethylene bags. Papaya seeds need air circulation during storage and can maintain viability for two years in paper bags at 10°C without significant loss in vigor (Doijode, 1993).

Seed storage with desiccants. High seed moisture is injurious to storage life, and chemical desiccants are used to regulate moisture during storage. Seeds stored with desiccants maintain initial viability. Seed viability and vigor are preserved for four years in seeds stored with silica gel, as compared to one year without silica gel under ambient conditions (see Figure 4.3) (Doijode, 1996). Viability decreases to 50 percent in seeds without silica gel after nine months of storage (Arumugam and Shanmugavelu, 1977).

FIGURE 4.2. Effect of Temperatures on Seed Viability in Papaya

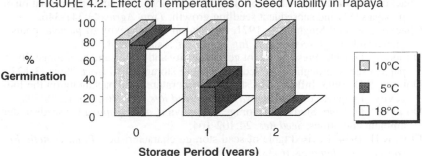

FIGURE 4.3. Seed Storage with Silica Gel in Papaya

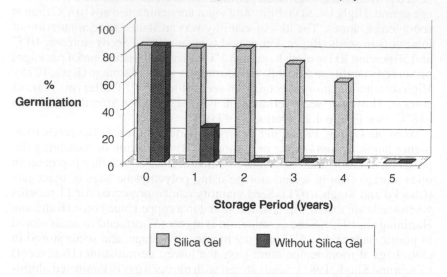

Storage Period (years)

□ Silica Gel   ■ Without Silica Gel

REFERENCES

Allan, P. 1969. Effect of seeds on fruit weight in *Carica papaya*. *Agroplantae* 1:163-170.

Arumugam, S. and Shanmugavelu, K.G. 1977. Studies on the viability of papaya seeds under different environments. *Seed Res.* 5:23-31.

Bass, L.N. 1975. Seed storage of *Carica papaya*. *HortSci.* 10:232.

Begum, H., Lavania, M.L., and Babu, G.H.V.R. 1988. Seed studies in papaya. II. Effects of presoaking treatments with gibberellic acids and thiourea on germination and vigor of aged seeds. *Seed Res.* 16:51-56.

Camejo, B. and Rivero, W. 1987. Study of germination and growth of pawpaw *(Carica papaya)* seedlings cv. Maradol in covered and uncovered nurseries. *Ciencia y Tecnica en la Agricultura Çitricos y Otros. Frutales* 10:85-93.

Chacko, E.K. and Singh, R.N. 1966. The effect of gibberellic acid on germination of papaya seeds and subsequent seedling growth. *Trop. Agric.* 43:341-346.

Chacko, E.K. and Singh, R.N. 1971. Studies on the longevity of papaya, phalsa, guava and mango seeds. *Proc. Int. Seed Test. Assoc.* 36:147-158.

Doijode, S.D. 1992. Preservation of germinability and enhancement of longevity in papaya seeds through controlled atmosphere storage. *Acta Hort.* 296:123-127.

Doijode, S.D. 1993. Influence of storage temperatures and packaging on the longevity of papaya *(Carica papaya)* seeds. *Seed Res.* Spl. Vol. No. 1:288-293.

Doijode, S.D. 1996. Studies on storage of papaya *(Carica papaya* L.) seeds under ambient conditions. *Seed Res.* 24:102-104.

Ellis, R.H. 1984. Revised table of seed storage characteristics. *Plant Genetic Resources Newsletter* 58:16-33.

Furutani, S.C. and Nagao, M.A. 1987. Influence of temperature, KNO_3, GA_3, and seed drying on emergence of papaya seedlings. *Scientia Hort.* 32:67-72.

Hewitt, E.J. 1963. The essential elements: Requirements and interactions in plants. In F.C. Steward (Ed.). New York: Academic Press, pp. 137-329.

Kalie, M.B. and Hartiningsih, M.G. 1973. Pawpaw seed storage experiment. *Bull. Penelitian Horti-Kultura* 1:13-17.

Koyamu, K. 1951. A preliminary note on the germination of papaya seeds. *Madras Agril. J.* 38:348-349.

Lange, A.H. 1961a. The effect of temperature and photoperiod on the growth of *Carica papaya. Ecology* 42:481-486.

Lange, A.H. 1961b. Effect of the sarcotesta on germination of *Carica papaya. Bot. Gaz.* 122: 305-311.

Limadiaz, S., Limadiaz, I., Valenzuela, Galindo R., and Macias, P. 1985. Study of seed viability of *Carica papaya* cv. Maradol Roza. *Centro Agricola* 12:119-130.

Makhija, M. and Jindal, P.C. 1983. Effect of different soil salinity levels on seed germination and seedling growth in papaya *(Carica papaya). Seed Res.* 11:125-128.

Mosqueda, V.R. 1969. The effect of various treatments on pawpaw seed germination. *Agricultura Technica en Mexico* 2:487-491.

Nagao, M.A. and Furutani, S.C. 1986. Improving germination of papaya seeds by density separation, KNO3 and GA. *HortSci.* 21:1439-1440.

Palaniswamy, V. and Ramamoorthy, K. 1987. Seed germination studies in papaya. *Prog. Hort.* 19:253-255.

Passera, C. and Spettoli, P. 1981. Chemical composition of papaya seeds. *Qualitas Plantarum* 31:77-83.

Perez, A., Reyes, M.N., and Cuevas, J. 1980. Germination of two papaya varieties: Effect of seed aeration, K-treatment, removing of the sarcotesta, high temperature, soaking in distilled H_2O and age of seeds. *J. Agril. University, Puerto Rico* 64:173-180.

Reyes, M.N., Perez, A., and Cuevas, J. 1980. Detecting endogenous growth regulators on the sarcotesta, sclerotesta, endosperm and embryo by paper chromatography on fresh and old seeds of two papaya varieties. *J. Agril. University, Puerto Rico* 64:164-172.

Saito, Y. and Yamamota, S. 1965. Some studies on the cultivation of tropical horticultural crops in the warm region of Japan. *Bull. Fac. Agric. Miyazaki* 11:206-219.

Salazar-Castro, R. 1982. Effect of the propagation medium on the germination and development of pawpaw seedlings. *Revista Instituto Colombiano. Agropecuario* 17:101-108.

Singh, R.M. and Singh, I.D. 1979. Note on genotypic response to seed size in papaya *(Carica papaya* L.). *Seed Res.* 7:37-40.

Singh, R.M. and Singh, I.D. 1981. Effect of methods and duration of storage on seed germination and seedling vigor in papaya *(Carica papaya* L.). *Seed Res.* 9: 67-72.

Storey, W.B. 1969. Papaya. In *Outlines of perennial crop breeding in the tropics,* Ferwada, F.P. and Wit, F. (Eds.), Misc. Paper 4. Wageningen: Landbonwhogeschool, p. 391.

Vecchio, V. and Shirwa, A.H. 1987. The effect of different methods of drying pawpaw seeds on their germination. *Revista di Agricoltura Subtropicale e Tropicale* 81:175-184.

Yahiro, M. 1979. Effects of seed pretreatment on the promotion of germination in papaya *(Carica papaya)*. *Memoirs of Fac. Agril. Kagoshima University* 15: 49-54.

5

Grape: *Vitis vinifera* L.

Introduction

Grape is a popular fruit crop in temperate regions. Its cultivation is expanding rapidly in the tropics and subtropics. It is grown for the juicy edible berries. Fruits are tasty and nourishing, containing high carbohydrates and minerals such as magnesium, potassium, phosphorus, and vitamins B_1 and B_2. Grape is easily digestible because it contains high amounts of glucose and fructose. It is also used in the preparation of juice, wine, syrup, and raisins.

Origin and Distribution

Grape has its origin in Central Asia, particularly in Armenia, and wild relatives are grown in areas between the Black and Caspian Seas. Grape is predominantly cultivated in Afghanistan, Australia, Canada, England, France, Germany, India, Israel, Italy, the Philippines, Portugal, Russia, Spain, and the United States.

Morphology

Grape belongs to family Vitaceae and has somatic chromosome number $2n = 38$. It is a dicotyledonous, deciduous, and woody climbing vine. Leaves are broad and cordate, with three to five lobes and intermittent tendrils. Flowers are borne in panicle formation and are mostly perfect. Fruit cluster is variable in size, form, and color; the fruit skin adheres to the pulp. Seeds are brown, none to four in number, and pear-shaped.

Seed Storage

Grape is propagated by cuttings, and seldom by seeds. However, seeds are used in evolving new cultivars through hybridization and in long-term conservation of genetic diversity through seed storage. Grape seeds show orthodox storage behavior, where drying or exposure to chilling temperatures does not kill seeds. Fresh seeds exhibit dormancy, which is beneficial for better seed storage.

Seed Collections

Grape is a cool-season crop. It grows well in temperate regions, except in areas of extreme cold. It needs warm growing conditions during fruit development and ripening. Rooted cuttings are used for grape propagation. Vines are suitably trained on arches, trellises, or poles (see Figure 5.1). Flowering occurs on new growth. Fruits are nonclimatric; therefore, ripe bunches are plucked for seed purposes. Berries are crushed without injuring the seeds. Seeds are separated from the pulp, washed in water, dried, and packed according to storage requirements.

Fruit maturity. Seeds from mature berries emerge sooner and produce more plantlets of higher dry weight than seeds from green berries. Grape berries harvested at the green and mature stages and stored in paper bags at 20°C for 12 months showed seed germination percentage of 14.5 and 73.5, respectively (Maeda, Pereira, and Terra, 1984).

Seed Germination

Freshly extracted seeds normally contain a high proportion of dormant seeds. Scott and Ink (1950) reported that fresh seeds had low germinability and low seedling vigor. Plants were dwarfs when such seeds were sown.

FIGURE 5.1. Grapevine with Ripe Berries

Seeds obtained from self-pollination are lower in viability than seeds from open pollination (Mamarov, Ivanov, and Katherov, 1958). Similarly, seeds obtained from grafted plants showed higher viability than seeds from self-rooted plants (Ivanov, 1961). Harmon and Weinberger (1959) noted that stratification or moist ripening results in higher seed germination. Grape seeds stored at 4 to 5°C for 60 days showed improved percentage of germination and early emergence (Forlani and Coppola, 1977), while Eris (1976) recommends storage for 120 days. Seed germination can also be hastened by keeping seeds in a peat-sand (1:1) mixture in polyethylene bags at 0 to 2°C (Misic, Lekic, and Todorovic, 1978). In another study, temperature regime (0 to 10°C) and storage duration (up to five months) increased germination. The most effective temperature for higher germination (40 percent) was 0°C. The effect decreased progressively with higher temperatures, and only 5 percent germinated at 10°C (Balthazard, 1974). However, a longer period of stratification at 1 to 3°C for 180 to 210 days improved germination (Dolgova, 1959). These seedlings grew vigorously and showed improved winterhardiness. Selim and colleagues (1981) reported that warm stratification (18°C for 30 or 60 days) followed by cold stratification (5°C for 30 days) gave higher germination and a higher number of normal seedlings. The improvement was attributed to an increase in endogenous level of gibberellic acid (GA)-like substances. Furthermore, gibberellic acid treatment replaces the chilling or light requirement for germination (Pereira and Maeda, 1986). A long prechilling treatment is rarely successful in promoting germination of dormant seeds. Gibberellins are more effective in promoting germination of dormant seeds than other growth substances (Kachru, Singh, and Yadav, 1972). Chohan and Dhillon (1976) reported that auxin activity increases with longer stratification. Although no gibberellin activity occurs in fresh seeds, it is higher in extracts of moist seeds stratified for 60 days. Abscisic acid-like inhibitor content is initially higher but is reduced on stratification. In many grape varieties, seeds are empty and do not germinate in spite of moist ripening or gibberellin applications (Chadha and Randhawa, 1967).

Heat treatment. Grape seed coat is hard and takes longer to imbibe. Immersing seeds in hot water (27 to 54°C) for 24 h also does not break the dormancy (Manivel and Weaver, 1974).

Light. Seed exposure to continuous light promotes the germination process. Pereira and Maeda (1986) reported that seeds of cv. Patricia germinated well in continuous light for 48 h but failed in dark conditions.

Storage Conditions

Grape seeds are short-lived under ambient conditions. Seeds deteriorate at higher temperature and relative humidity. Dolgova (1959) noted poor germination of seeds when stored in dry conditions for six to eight months. Viability is preserved when seeds are stored in the berry itself or stratified in damp sand at 1 to 3°C for 180 to 210 days. Seed drying after harvest and/or stratification at 18.3 to 21.1°C reduces germination slightly, and even more through drying without subsequent stratification. Dry seeds stored at 4.4°C for three months showed higher germination than those stored at room temperatures (Singh, 1961). Seeds stored in paper or polyethylene bags at 20°C germinated poorly, unlike those stored at 2°C in polyethylene bags (Eynard, Gay, and Savino, 1972). Furthermore, fresh seeds stored in glass jars or paper bags remained viable at 10°C, and dry seeds, at 20°C (Maeda, Pereira, and Terra, 1985). Seeds stored in polyethylene bags at 5 and −18°C for two years resulted in 45 and 41 percent germination, respectively, after gibberellic acid (2000 ppm) applications (Doijode, 1993). Viability (67 percent) was extended further, for five years, by packing seeds in polyethylene bags and storing at 5°C (Doijode, unpublished data).

REFERENCES

Balthazard, J. 1974. The effect of temperature on true and apparent dormancy in grape vine seeds. *Comptes Rendus Hebdomadaires des Seances del Acad. des Sci. D.* 278:2143-2144.

Chadha, K.L. and Randhawa, G.S. 1967. Studies on grape seed germination: A review. *Indian J. Hort.* 24:181-187.

Chohan, G.S. and Dhillon, B.S. 1976. Seed dormancy and endogenous growth substances in Anab-e-Shahi grapes. *Vitis* 15:5-10.

Doijode, S.D. 1993. Storage of grape seeds and improvement of their germinability with certain chemical treatments. *J. Mah. Agril. Universities* 18:107-109.

Dolgova, A.E. 1959. The effect of storage conditions of grape seeds on germination and growth of the seedlings. *Sad i Ogorod* No. 9:65-68.

Eris, A. 1976. On the influence of growth regulators and stratification on germination of seeds of the vine cv. Muscat Hamburg. *Mitteilungen Rebe Wein Obstbau Fruchteverwertung* 26:85-90.

Eynard, I., Gay, G., and Savino, P.G. 1972. Studies on the germinability of grape vine seeds as affected by the method of storage, physical and chemical seed treatment and ambient conditions during germination. *Ann.della Facolta di Sci. Agranedella University degli Stud di Torina* 8:127-156.

Forlani, M. and Coppola, V. 1977. The effects of cold storage, gibberellic acid, and temperature on the germination of *Vitis vinifera* seeds. *Revista di Viticoltura di Enologia* 30:445-451.

Harmon, F.N. and Weinberger, J.H. 1959. Effect of storage and stratification on germination of *Vinifera* grape seeds. *Proc. Am. Soc. Hort. Sci.* 73:147-150.

Ivanov, I.K. 1961. Characteristics of seeds and seedlings obtained from self rooted and grafted grape vines. *Agrobiologija* 6:814-820.

Kachru, R.B., Singh, R.N., and Yadav, I.S. 1972. Physiological studies on dormancy in grape seeds *(Vitis vinifera).* II. On the effect of exogenous application of growth substances, low chilling temperature and subjection of the seed to running water. *Vitis* 11:289-295.

Maeda, J.A., Pereira, M.De.F.D.A., and Terra, M.M. 1984. Effect of the fruit developmental stage on seed quality of the grape vine cultivar Patricia. *Bragantia* 43:659-666.

Maeda, J.A., Pereira, M.De.F.D.A., and Terra, M.M. 1985. Storage conditions and grape vine seed viability and dormancy. *Bragantia* 44:245-254.

Mamarov, P., Ivanov, J., and Katherov, K. 1958. Effect of presowing treatments of vine seeds on germination. *Sborn CSI Akad. Zemed Ved Rostt Vyroba* 31:1211-1216.

Manivel, L. and Weaver, R.J. 1974. Effect of growth regulators and heat treatment on germination of Tokay grape seeds. *Vitis* 12:286-290.

Misic, P., Lekic, N., and Todorovic, R. 1978. Studies on methods for improving seed germination in grape vine. *Savremena Poljoprivreda* 26:77-80.

Pereira, M.De.F.D.A. and Maeda, J.A. 1986. Environmental and endogenous control of germination of *Vitis vinifera* seeds. *Seed Sci. Technol.* 14:227-235.

Scott, D.H. and Ink, D.P. 1950. Grape seed germination experiments. *Proc. Am. Soc. Hort. Sci.* 56:134-139.

Selim, H.H., Ibrahim, F.A., Fayek, M.A.E.L., El-Dean, S.A.S., and Gamel, N.M. 1981. Effect of different treatments on germination of Romi Red grape seeds. *Vitis* 20:115-121.

Singh, S.N. 1961. Germination of grape hybrid seeds by chilling. *Curr. Sci.* 30:62.

– 6 –

Pineapple:
Ananas comosus (L.) Merr.

Introduction

Pineapple is a major tropical fruit cultivated in hot and humid regions for the edible fruits. Fruits are eaten fresh after peeling the rind. It is also canned in pieces or processed for jams, juices, and squashes. Fruits are a rich source of vitamins A, B, and C and of minerals such as calcium, phosphorus, and iron. They are a good source of the protease enzyme bromelain, which is used in medicine and in the food-processing industry.

Origin and Distribution

Pineapple originated in Brazil. Many primitive and wild relatives are found in the Caribbean, Central America, and northern South America. Its cultivation later spread to other countries, such as Australia, China, Colombia, Cuba, Ghana, India, Indonesia, Japan, Kenya, Malaysia, Mexico, Mozambique, Nigeria, the Philippines, South Africa, Taiwan, Thailand, and the United States.

Morphology

Pineapple belongs to family Bromeliaceae and has chromosome number 2n = 50. The plant, a perennial and monocotyledonous herb, grows up to 1 m in height. Leaves are narrow, thick, rosette shaped, with a pointed tip and spiny margin, and are about 60 to 80 in number. The inflorescence is compact and bears 100 to 200 flowers. Flowers are perfect, opening first at the basal portion and later moving upward to produce fruit parthenocarpically. Fruit is cylindrical and dark green, changing to yellow or reddish orange on ripening. Fruit is seedless and has yellow flesh. Artificial pollination produces 2,000 to 3,000 seeds. Seeds are brown and rough; embryos are small with hard endosperm.

Seed Storage

Pineapple is commonly propagated by vegetative means. Shoots from crown, slips, and suckers are used for propagation. Seeds are used in evolving new cultivars through hybridization. Pineapple seeds show orthodox storage behavior. They can withstand desiccation, and longevity is extended at lower-temperature storage. Many cultivars do not produce seeds, and fruit develops parthenocarpically. In certain cases, seeds are produced through artificial pollination. These can be used to a certain extent for conservation of genetic diversity.

Seed Collections

Pineapple is a warm-season crop. It needs hot and dry conditions for rapid growth. Fruit takes longer to grow at higher latitudes, and fruit quality improves at lower temperatures. It requires deep, fertile, and sandy loam soils. Plants take longer (four years) to bear fruit when propagated by seeds than when grown from suckers (18 months). Fruit becomes yellow to reddish orange on ripening. Seed sets on artificial pollination and takes about five to six months for maturation. Seeds are present inside the carpel about 1.5 cm deep and are extracted by cutting fruit longitudinally. These are washed, dried, and used for storage.

Seed Germination

Seeds have a very tough seed coat and hard flinty endosperm. Therefore, they germinate slowly and unevenly (Collins, 1968). Normally, it takes ten days for germination after acid scarification at 30°C. Seed incubation at 30 to 35°C and scarification with concentrated sulfuric acid for 30 min is beneficial for increased germination (Gopinomy, Balkrishnan, and Kannan, 1976). Iyer, Singh, and Subramanyam (1978) reported faster and higher germination after exposing seeds to intermittent mist (for 20 seconds [sec]/ 3.5 min intervals). This is attributed to softening of the hard seed coat or leaching of water-soluble inhibitors by the water spray.

Storage Conditions

Pineapple seeds are orthodox in storage behavior. They store better under cool, dry conditions. Well-dried seeds are suitably packed in moisture-proof containers and stored at low temperature for longer storage.

REFERENCES

Collins, J.L. 1968. *The pineapple.* London: Leonard Hill.

Gopinomy, R., Balkrishnan, S., and Kannan, K. 1976. A note on germinating seeds of pineapple *(Ananas comosus). Agril. Res. J. Kerala* 14:194-195.

Iyer, C.P.A., Singh, R., and Subramanyam, M.D. 1978. A simple method for rapid germination of pineapple seeds. *Scientia Hort.* 8:39-41.

Guava: *Psidium guajava* L.

Introduction

Guava is an important fruit crop in the tropics and subtropics. Mature green or ripe fruits are consumed fresh and used in preparation of products such as juices, pastes, jams, jellies, and pies. Fruits are rich in pectin and vitamin C, and red-fleshed cultivars are high in carotene content. Leaves and barks contain a large amount of tannin and are used in the dyeing and tanning industry. Leaves have medicinal properties and are used in controlling diarrhea and stomach ailments.

Origin and Distribution

Guava's wild relatives and primitive types are found in Brazil, Mexico, and Peru. Later its cultivation spread to other African and Asian countries. Guava is cultivated predominantly in Algeria, Australia, Brazil, China, Colombia, Cuba, Egypt, India, Israel, Mexico, Peru, the Philippines, South Africa, the United States, and the West Indies.

Morphology

Guava belongs to family Myrtaceae and has chromosome number 2n = 22. It is a small, dicotyledonous, semi-deciduous tree growing up to 3 to 10 m in height. Foliage is thick and dark green. Leaves are opposite and oblong with a rough surface. Flowers are white, with one to three in a cluster, axillary, hermaphrodite, and fragrant. Calyx is entire in bud and splits into four to six lobes; petals number four to five; stamens are numerous; the ovary has four to five locules. Fruit is a berry, globular or pear-shaped; the exocarp is pale green to bright yellow; the mesocarp is white, yellow, pink, or red and fleshy and is embedded with many seeds. Seeds are small (3 to 5 mm), hard, reniform, and yellowish in color.

Seed Storage

Guava is propagated both sexually and asexually. Many existing orchards have been raised from seedlings. Guava seeds show orthodox storage behavior (Becwar, Stanwood, and Leonhardt, 1983). They preserve high viability and vigor under dry, cool conditions. Seeds can be used for long-term conservation of genetic diversity in the gene bank (Ellis, Hong, and Roberts, 1985). Seeds are also involved in crop improvement through hybridization programs. Guava seeds can be stored well at low moisture and low temperature without significant loss of viability (Ellis, 1984).

Seed Collections

Guava is cultivated in a wide range of climatic and soil conditions. It prefers warm and dry conditions for optimum growth. Frost and temperatures below 16°C affect plant growth. It grows well in deep, friable, well-drained soils and is tolerant to saline conditions. Seeds and air layers are used for propagation. Air layer plants do not show variations and retain the identity of the mother plant. Tree commences flowering from the second year onward. Both self- and cross-pollination by insects occur. Fruit takes about five months to mature following anthesis. Healthy and ripe fruits are selected for seed purposes; seeds are separated from the pulp, washed thoroughly, dried in shade, and packed suitably for storage.

Seed Germination

Guava seeds germinate poorly and unevenly and require more time for seedling emergence. Teaotia and Singh (1972) reported a variation in germination percentage from 22 to 72 in different cultivars. Poor germination of guava seeds is primarily attributed to dormancy (Teng and Hor, 1976) and to hard testae (Singh and Soni, 1974). Seed scarification, either mechanically or chemically with acids, overcomes delay and the slow germination process. Seed soaking in tepid water before sowing increases the germination percentage, and use of cold water is safer than using hot water (Haq, Khan, and Faridullah, 1973). Seed soaking in cold water for 12 h (Singh and Soni, 1974) or in boiling water for 5 min (Sinha, Verma, and Koranga, 1973) is beneficial for higher germination. Seed soaking in concentrated hydrochloric acid for 3 min (Singh and Soni, 1974) or in concentrated sulfuric acid for 5 min (Sinha, Verma, and Koranga, 1973) eliminates the dormancy. Further soaking of seeds in ethephon (5,000 ppm) increased seedling emergence and encouraged better vegetative growth of plants (Sinha, Verma, and Koranga, 1973). Exposure of seeds to alternate temperatures of 20/30°C for 16/8 h, respectively, improved seed germination (Ellis, Hong, Roberts, 1985).

Storage Conditions

Guava seeds show orthodox storage behavior. Thus, well-dried seeds appear to preserve longer at low temperature. These seeds retain viability for about one year under ambient conditions. According to Chacko and Singh (1971), seed viability reduced significantly after 18 months of storage when stored with or without charcoal in open or sealed containers.

REFERENCES

Becwar, M.R., Stanwood, P.C., and Leonhardt, K.W. 1983. Dehydration effects on freezing characteristics and survival in liquid nitrogen of desiccation-tolerant and desiccation-sensitive seeds. *J. Am. Soc. Hort. Sci.* 108:613-618.

Chacko, E.K. and Singh, R.N. 1971. Studies on the longevity of papaya, phalsa, guava and mango seeds. *Proc. Inter. Seed Test. Assoc.* 36:147-158.

Ellis, R.H. 1984. Revised table of seed storage characteristics. *Plant Genetic Resources Newsletter* 58:16-33.

Ellis, R.H., Hong, T.D., and Roberts, E.H. 1985. *Handbook of seed technology,* Volume II. Rome: International Board for Plant Genetic Resources, pp. 511-513.

Haq, F., Khan, M.S., and Faridullah, I. 1973. Germination trial on guava seeds. *J. Agril. Res. Pakistan* 11:121.

Singh, S. and Soni, S.L. 1974. Effect of water and acid soaking periods on seed germination in guava. *Punjab Hort. J.* 14:122-124.

Sinha, M.M., Verma, J.P., and Koranga, D.S. 1973. Studies on the seed germination of guava (*Psidium guavajava* L.). I. Effect of scarification and plant growth regulator treatments. *Prog. Hort.* 5:37-40.

Teaotia, S.S. and Singh, R.D. 1972. Standardization of rootstocks of guava. I. Studies on seed germination, congeniality and vigor of various guava species and varieties. *Prog. Hort.* 4:23-34.

Teng, Y.T. and Hor, Y.L. 1976. Storage of tropical fruits seeds. In *Seed technology in the tropics,* Chin, H.F., Enoch, I.C., and Raja Harun, R.M. (Eds.), Malaysia: University Pertanian, pp. 135-146.

Banana (*Musa* spp.)

Introduction

Banana is a leading fruit crop of the tropics. It is cultivated for its edible fruits. Ripe fruits are consumed fresh, while unripe fruits are cooked as a vegetable. Almost all parts of banana plants are used in one form or another, such as for wrappers, fibers, in medicine, in religious functions, as cattle feed, etc. Bananas are a rich source of energy, and they contain high amounts of carbohydrates, calcium, phosphorus, iron, magnesium, sodium, potassium, and vitamins A, B, and C.

Origin and Distribution

Banana originated in Southeast Asia. The cultivated banana is derived from *Musa acuminata* (A genome) and *M. bulbisiana* (B genome) types. Banana grows well in humid tropical lowlands and is predominantly distributed between 30°N and 30°S of the equator. Major banana-growing countries are Australia, Bangladesh, Brazil, Burundi, Colombia, Cuba, Ecuador, Egypt, Fiji, Ghana, Guatemala, India, Indonesia, Israel, the Ivory Coast, Kenya, Malaysia, Mexico, Panama, the Philippines, South Africa, Tanzania, Uganda, the United States, Venezuela, the West Indies, and Zaire.

Morphology

Banana belongs to family Musaceae and has chromosome numbers 2n = 22, 33, and 44. Seed sets in wild diploid banana but fails in triploids due to sterility. Banana is a perennial, monocotyledonous, robust herb. The corm, or underground stem, has short internodes. It takes 7 to 9 months for flowering after planting suckers. The inflorescence is a spike that produces 12 to 20 flowers per node. The first 5 to15 basal nodes produce female flowers, while the terminal node forms male flowers (see Figure 8.1). Fruit is a berry and takes 90 days for maturity from anthesis. Seeds are subglobose or angu-

FIGURE 8.1. Banana Fruit Bunch Supported by a Pole

lar, 5 mm in diameter, very hard, and endospermic, with a minute embryo at micropylar end.

Seed Storage

Banana is propagated by both sexual and asexual methods. Seeds are used in hybridization for evolving new cultivars, while suckers are planted for raising commercial orchards. Banana seeds show orthodox storage behavior, retaining high viability on drying and conservation at chilling temperatures.

Seed Collections

Banana is a warm-season crop. It requires a hot, humid climate for optimum growth. Heavy winds and hailstorms damage the crop. Plant needs support, especially during fruit development. It is a shallow-rooted plant and requires fertile, well-drained soil. It is a cross-pollinated crop, and pollination is achieved by ants, bees, wasps, and bats. Fruit matures in 11 to 16 months after planting, depending on cultivars and growing conditions. On maturity, bunches are harvested and fruits allowed to ripen. Seeds are extracted from the pulp, washed, dried, and packed for storage.

Seed Germination

Banana seeds germinate readily after extraction. However, they become dormant on drying. Seed germination is hypogeal. Seeds germinate with water imbibitions, and a micropylar plug extrudes, followed by the radicle, which is short-lived and quickly replaced by seminal adventitious roots. Then, the plumule grows and becomes a normal seedling. Seed germination is higher in seeds extracted from matured fruits than in those from immature fruits. Low and erratic germination were attributed to degree of seed maturity (Simmonds, 1959). Seed viability is decided by climatic conditions, period of pollination, and position of seeds in fruit. Banana seeds germinate between temperatures of 10 to 37°C, and germination is affected beyond this range of temperature (Stotzky and Cox, 1962). Dormant seeds do not germinate in this temperature regime. However, they do germinate readily at alternate temperatures of 15/35°C. Chipping imbibed seeds gives rapid germination. Nondormant seeds germinated at 25 to 35°C within three weeks of sowing. Storage of *Musa bulbisiana* seeds in a desiccator at room temperature for three months resulted in an increase in germination from 73 to 95 percent (Simmonds, 1952). Seed germination can be enhanced by six months' storage in 5 percent carbon dioxide atmosphere (Purseglove, 1972).

Storage Conditions

Banana seeds are orthodox in storage behavior and germinate readily. Seeds withstand desiccation and their longevity is extended at low temperature. Purseglove (1972) reported that seeds buried in soil remained dormant for more than one year under dry conditions. Dried seeds remain viable for a few months to two years (Chin, 1996). Seeds are stored suitably in moisture-proof containers, such as thick polyethylene bags, and in laminated aluminum foil pouches at low or subzero (–20°C) temperatures, or even at cryogenic temperatures, for longer conservation of viability and vigor.

REFERENCES

Chin, H.F. 1996. Germination and storage of banana seeds. In *Proc. New Front. in resistance breeding for nematode, Fusarium and Sigatoka, Kuala Lumpur.*
Frison, E.A. and Horry, J.P. (Eds.). Univ. Pertanian Serdang Selangor, Malaysia, pp. 218-229.
Purseglove, J.W. 1972. *Tropical crops monocotyledons.* London: Longman, p. 361.
Simmonds, N.W. 1952. The germination of banana seeds. *Trop. Agril. Trinidad* 29:2-16.
Simmonds, N.W. 1959. Experiments on the germination of banana seeds. *Trop. Agril.* 36:259-274.
Stotzky, C. and Cox, E.A. 1962. Seed germination studies in *Musa*. II. Alternating temperature requirement for the germination of *Musa bulbisiana. Amer. J. Bot.* 49:763-770.

– 9 –

Jackfruit:
Artocarpus heterophyllus Lam.

Introduction

Jackfruit is largely grown in the tropics and subtropics. Normally, a few trees are planted in orchards and home gardens. Fruits are appreciated for their sweet crisp or soft pulp and characteristic flavor. Immature fruits are cooked as vegetables. Pulp of ripe fruits is eaten fresh, cooked, or preserved in syrup. Seeds are boiled, roasted, and eaten. The green foliage and rind are used as cattle feed, and the wood from the plant forms an excellent timber. The leaves also possess medicinal properties, and the fruits are a rich source of iron, phosphorus, calcium, and vitamin A.

Origin and Distribution

Jackfruit originated in India. The wild relatives are found in western parts of India. Later, it was introduced to other countries in Africa and Asia. Jackfruit is cultivated in Brazil, India, Malaysia, Myanmar, the Philippines, Sri Lanka, Thailand, and the West Indies.

Morphology

Jackfruit is a member of the Moraceae family and has chromosome number 2n = 56. It is an evergreen, monoecious, dicotyledonous large tree that grows up to 20 m in height. All parts of the plant produce white latex. Foliage is dark green; leaves are alternate and ovate, with shiny upper surfaces. Male and female inflorescences are borne separately, and the male inflorescence is axillary on the leaf twigs, 15 to 30 centimeters (cm) long, drooping, green, and club-shaped. The female inflorescence forms on short stout twigs on the trunk and thicker branches, and it is oblong in shape. Fruit is a large syncarp, dark green, pear-shaped, and glabrous with fleshy spines, and

73

it contains many seeds. Pulp is yellow and waxy. Seeds are large and oblong with a thick gelatinous covering.

Seed Storage

Jackfruit is commonly propagated from seeds sown directly in the field or from transplanted seedlings. Fruits contain normally around 100 seeds in table types and 500 seeds in inferior types. Seeds weighing between 3 to 6 g are considered good for planting, and those less than 3 g are not suited for propagation. Seeds must be sown immediately after extraction from fruit, or they lose their viability. Jackfruit seeds exhibit recalcitrant storage behavior, where seeds are killed on drying and exposure to chilling temperatures. Such seeds are difficult to store for long periods. However, seeds can be stored for a few weeks to few months with available storage methods.

Seed Collections

Jackfruit grows well in hot, humid climates. It prefers deep, fertile, and well-drained soil. Seeds, which are predominantly used in propagation of jackfruit, germinate up to 15 days after extraction, and thereafter germination declines. The flowers are pollinated by insects and take 100 to 120 days for fruit maturity. In general, a tree bears 200 to 300 fruits (see Figure 9.1). Fruits are harvested at maturity and stored for a few days for ripening. They emit a peculiar odor on ripening. Large, healthy, and normally shaped fruits are selected for their seeds. Seeds are extracted and washed, with large and bold seeds selected for storage.

Seed Germination

Fresh seeds germinate readily and take three to eight weeks to complete the germination. Dry and shriveled cotyledonous seeds do not germinate. Viviparous seed germination is reported in early maturing types (25 to 36 percent) and in late-maturing types (37 to 39 percent) (Maurya, 1987). Seed germination is higher at a constant temperature range of 25 to 30°C and in the presence of light (Ellis, Hong, and Roberts, 1985). It is improved further on seed soaking in gibberellic acid (100 ppm) (Shanmugavelu, 1970; 1971) and naphthaleneacetic acid (25 ppm) (Sinha and Sinha, 1968).

Storage Conditions

Jackfruit seeds are short-lived under ambient conditions. Their vigor decreases on increase in storage period. They lose viability and vigor rapidly with decrease in seed moisture content. High-moisture seeds failed to preserve viability at low temperature and showed chilling injury and loss of vi-

FIGURE 9.1. Jackfruit Tree with Many Fruits

ability. Fresh seeds gave rapid and higher germination and better seedling survival than sowing the seeds stored at 6°C for 15 days (Chiesotsu, Kar, and Sanyal, 1995). Jackfruit seeds stored at ambient temperatures and at 6°C germinated well, in the range of 80 to 86 percent, after 22 days of storage; none of the seeds germinated after 38 days of storage (Panggabean, 1979). Moist storage in a screw cap bottle at 16 to18°C is an ideal method that preserves seed viability for 45 days (Fernandez, 1982). Tang and Fu (1993) preserved high seed viability and vigor for more than two years by packing washed seeds in perforated polyethylene bags and storing them at 15°C. Seed storage at 5 and 10°C showed chilling injury after 40 and 300 days of storage, respectively.

REFERENCES

Chiesotsu, S., Kar, P.L., and Sanyal, D. 1995. A note on germination and seedling vigor of jackfruit seeds as influenced by growth regulators and storage. *Hort. J.* 8:151-155.

Ellis, R.H., Hong, T.D., and Roberts, E.H. 1985. *Handbook of seed technology for gene banks,* Volume II. Rome: International Board for Plant Genetic Resources, p. 502.

Fernandez, P.G. 1982. Effect of air drying, moist storage, and soaking on the physiology of jackfruit *(Artocarpus heterophyllus)* seeds. *Laguna Coll. Tech. Bull.* p. 145.

Maurya, K.R. 1987. A note on viviparous seed germination in jackfruit. *Prog. Hort.* 19:137-138.

Panggabean, G. 1979. The effect of storage on the germination of jackfruit seeds *(Artocarpus heterophyllus)*. *Bulletin Penelitian Hortikultura* 7:39-42.

Shanmugavelu, K.G. 1970. Effect of gibberellic acid on seed germination and development of seedlings of some tree plant species. *Madras Agric. J.* 57:311-314.

Shanmugavelu, K.G. 1971. Effect of growth regulators on jack. *Madras Agric. J.* 58:97-103.

Sinha, M.M. and Sinha, S.N. 1968. Effect of naphthalene-alpha-acetic acid on the germination of jack seeds. *Sci. Cult.* 34:372-373.

Tang, L.E. and Fu, J.R. 1993. A reasonable suggestion on moist storage of jackfruit *(Artocarpus heterophyllus)* seeds. *Acta Scientiarum Natu.University Sunyatseni* 32:111-115.

10

Pomegranate: *Punica granatum* L.

Introduction

Pomegranate is a popular dessert fruit in the tropics and subtropics. It is extensively cultivated in the Mediterranean region for the fruit's juicy pulp. It is also grown in home gardens as an ornamental plant. The fruit is processed for juice, syrup, and jelly preparations. Seeds are used in preparation of several food dishes. It is rich in minerals such as calcium, iron, phosphorus, and vitamins B and C. The bark from the plant is used for medicinal purposes, such as curing stomach ailments and diarrhea, and the seeds show antibacterial properties.

Origin and Distribution

Pomegranate is native to Iran. Its wild types and landraces are grown in Central Asia, including in Afghanistan, Iran, and Pakistan. Pomegranate is cultivated in Afghanistan, Bulgaria, China, Egypt, India, Indonesia, Iran, Israel, Italy, Japan, Morocco, Myanmar, Pakistan, Russia, Spain, Syria, Tunisia, Turkey, and the United States.

Morphology

Pomegranate belongs to family Punicaceae and has chromosome number $2n = 16$. It is a small, bushy, dicotyledonous tree that reaches 2 to 4 m in height. Leaves are opposite, dark green, glossy, and deciduous in cooler climates. Flowers are showy, orange-red, and borne in clusters; the calyx is campanulate with five to seven lobes; petals number five to seven and are wrinkled; stamens are numerous; the ovary has three to seven cells. Fruit is a berry, leathery skinned, globose or spherical, brownish yellow to red, and surrounded by persistent calyx. Interior of fruit is composed of chambers that are filled with numerous seeds. Seeds are about 1 to 1.5 mm long and are surrounded by pink juicy pulp.

Seed Storage

Earlier, many pomegranate orchards were raised from seedlings. These plants required a longer period to bear fruit and showed variations in fruit qualities. Of late, vegetative propagation by hard cutting of the wood has become a common practice in pomegranate cultivation. It gives uniform, early bearing, and true-to-type plants. However, seedlings are used in developing new cultivars through hybridization, and seeds are used in long-term conservation of genetic diversity in the gene bank. Seeds show orthodox storage behavior, withstand drying, and are capable of preserving viability at low temperatures.

Seed Collections

Pomegranate grows better in semiarid climates, where cool winters and hot dry summers are prevalent. It fruits poorly under very humid conditions and can withstand drought conditions. It needs a dry hot climate during fruit development for high-quality fruit. Deep loamy soils are ideal for cultivation. Plants flower throughout the year. Fruits take six months to ripe after anthesis. Self- and cross-pollination occur, the latter mainly by beetles. Plants commence bearing from the fourth year onward (see Figure 10.1). Healthy and mature fruits are selected for seed purposes. Seeds are extracted from fruit, washed in water, and surface dried. The fresh seeds contain about 70 to 75 percent moisture, which is reduced to 5 to 7 percent for storage.

Seed Germination

Pomegranate seeds germinate rapidly at constant temperatures of 30 to 35°C. Seeds germinated at 35°C give low-vigor seedlings. Seed exposure to alternate temperatures of 20/35°C for 16/8 h, respectively, promotes higher germination. The germination process is complete 28 days after sowing (Ellis, Hong, and Roberts, 1985). Seed germination is faster within 14 to 30 days of sowing when seeds are exposed to prechilling, 1 to 5°C for 30 days (Riley, 1981). According to Cervelli (1994), removal of the fleshy seed coat improves seedling emergence by 5 percent, and subsequent washing in water for 48 h increases it further, by 26 to 62 percent.

Storage Conditions

In pomegranate, the pericarp is woody and leathery, loses moisture slowly, and protects seeds from insects and diseases. Mature fruits are stored for five to six months in cool conditions (Anonymous, 1986). Subsequently, seeds are extracted and used for further storage or for propagation.

FIGURE 10.1. Early Fruiting in Pomegranate Tree

Seed viability is preserved well in cool, dry conditions. Fruits stored for a shorter period retain seed viability and vigor. Riley (1981) recommended drying of seeds to 70 percent of their harvest weight and storing them at 4 to 5°C to preserve seed viability for three years.

REFERENCES

Anonymous. 1986. *Genetic resources of tropical and sub-tropical fruits and nuts.* Rome: International Board for Plant Genetic Resources, pp. 97-99.

Cervelli, C. 1994. Effect of thermic treatments and seed manipulation on emergence of dwarf pomegranate (*Punica granatum* L.). *Acta Hort.* 362:189-195.

Ellis, R.H., Hong, T.D., and Roberts, E.H. 1985. *Handbook of seed technology for gene banks,* Volume II. Rome: International Board for Plant Genetic Resources, pp. 552-553.

Riley, J.M. 1981. Growing rare fruits from seeds. *Calif. Rare Fruit Grower Yearbook* 13:1-47.

11

Avocado: *Persea americana* Mill.

Introduction

Avocado is an important tropical fruit having high nutrient contents and the highest energy value among fruits. Fruits are normally consumed fresh along with sugar, cocoa, and other flavoring compounds. They are rich in unsaturated fats, proteins, minerals, and vitamins A, C, G, and thiamine. Avocado fruits, which are available throughout the year, have a nutlike flavor. Seeds also contain high amounts of oil and are used for cosmetic and medicinal purposes.

Origin and Distribution

Avocado originated in Central America. Later it spread to other tropical and subtropical regions. It is cultivated in Australia, Brazil, Chile, Cuba, Ecuador, Haiti, Indonesia, Israel, Kenya, Madagascar, Mexico, Peru, South Africa, Spain, Thailand, and the United States.

Morphology

Avocado belongs to family Lauraceae and has chromosome number $2n = 24$. Avocado is an evergreen tree with a short trunk, large branches, and a shallow root system. Leaves are simple, ovate, and spirally arranged. Inflorescence is compact and axillary with numerous yellow flowers. Flowers are perfect, protogynous, and fragrant. Fruit is a fleshy berry, pyriform or globose, and yellowish green to purple in color. Flesh has a butterlike taste and is yellow in color. Seeds are large, globose, single, and whitish or pink in color.

Seed Storage

Avocado is commonly propagated by vegetative means due to the heterogeneous nature of seeds. Many existing avocado orchards have been raised through seedlings. Seeds are sown immediately after extraction for raising rootstocks. These are short-lived and remain viable for two to three weeks after removal from fruit. Avocado seeds appear to be recalcitrant; they lose viability on desiccation and are susceptible to chilling temperatures.

Seed Collections

Avocado grows well in tropical conditions. Some cultivars are susceptible to cold and are damaged by strong winds. It grows well in deep, fertile, sandy loam soil. It is propagated by grafting. Plants show a biennial bearing habit. Avocado is a cross-pollinated crop, and pollination occurs mainly by bees. The percentage of fruit set is low. Healthy and bold seeds are separated from the fruit, washed, and used immediately or processed for short-term storage.

Seed Germination

Seed germination is hypogeal and higher at a constant temperature of 25 to 30°C. Seedling emergence is slow and nonuniform due to the hard seed coat. Removal of the seed coat promotes early germination. Halma and Frolich (1949) reported that cutting a small portion off both ends of the seed gives quicker and even germination. Later, Johnston and Frolich (1957) suggested that cutting a slice off the tip and base of the seeds gives higher and uniform germination. Seed soaking in gibberellic acid (125 to 1,000 ppm) resulted in higher germination and better seedling growth (Leal, Krezdorn, and Marte, 1977).

Storage Conditions

Avocado seeds are recalcitrant in storage behavior and remain viable for two to three weeks under ambient conditions. Seed viability can be maintained for several months by preserving seeds in sand or sawdust at 4.4 to 7.2°C (Johnston and Frolich, 1957), and for eight months in dry peat moss at 5.5°C (Halma and Frolich, 1949). Thin polyethylene (0.025 mm) bags are well suited for storage because they allow gaseous exchange and preserve viability comparatively longer. However, seeds stored in polyethylene bags with holes invite *Penicillium* or *Aspergillus* fungal species (Paulus, Nelson, and Zentmyer, 1976). Brokaw (1978) noted that periodic exchange of air in plastic bags and the use of thinner packaging help maintain high viability (90 percent) for six months. Similarly, Spalding, Knight, and Reeder (1976) preserved avocado seeds for five months at 4.4°C when sealed in polyethylene bags along with fungicides.

REFERENCES

Brokaw, W.H. 1978. Sub-tropical fruit tree production: Avocado as a case study combined. *Proc. Inter. Pl. Prop. Soc.* 27:113-121.

Halma, F.F. and Frolich, E. 1949. Storing avocado seeds and hastening germination. *Calif. Avocado Soc. Yearbook* 32:136-138.

Johnston, J.C. and Frolich, E.F. 1957. Avocado Propagation. *Circ. Calif. Agric. Exp. Stat.* 463:19.

Leal, F.J., Krezdorn, A.H., and Marte, R.J. 1977. The influence of gibberellic acid on the germination of avocado seeds. *Proc. Fl. St. Hort. Soc.* 89:258-261.

Paulus, A.O., Nelson, J., and Zentmyer, G.A. 1976. Avocado seed treatment and storage trial. *Calif. Avocado Soc. Yearbook* 59:70-71.

Spalding, D.H., Knight, R.J., Jr., and Reeder, W.F. 1976. Storage of avocado seeds. *Proc. Fl. St. Hort. Soc.* 89:257-258.

12

Litchi: *Litchi chinensis* Sonn.

Introduction

Litchi is a popular dessert fruit in Southeast Asia. The translucent juicy aril of the fruit is eaten fresh, canned, preserved in syrup, dried, or frozen. Litchi fruit dried and used as a nut is popular in China. Fruit contains high amounts of carbohydrates, calcium, phosphorus, and vitamin B. Seeds are used for medicinal purposes.

Origin and Distribution

Litchi belongs to family Sapindaceae and has chromosome number $2n = 28$. Litchi originated in southern China. Later its cultivation spread to other subtropical regions. It is largely cultivated in Australia, Bangladesh, Brazil, China, India, Indonesia, Israel, Japan, Madagascar, Mauritius, Mexico, New Zealand, Seychelles, South Africa, Taiwan, Thailand, the United States, Vietnam, and the West Indies.

Morphology

Litchi is an evergreen, dense, dicotyledonous tree that grows up to 10 m in height. Leaves are glossy, green, and pinnate with two to nine leaflets. Flowers are small, green, or yellow, in terminal panicles, with male and female flowers. Fruits are globose or oblong, borne in loose clusters with long peduncles, and the rind is dark red or yellow, brittle, and covered with sharp tubercles. Seeds are dark brown and covered with a pearl-white, translucent, juicy aril (see Figure 12.1).

Seed Storage

Litchi is propagated sexually and asexually. Seeds do not produce true-to-type plants and take longer to bear fruit, which is often of inferior quality. However, many old existing orchards have been raised from seedlings.

FIGURE 12.1. Litchi Fruits and Dark Recalcitrant Seeds

Presently air layering is a popular method for multiplication of litchi plants. Such plants bear fruit within four to six years of planting. Seeds are commonly used in raising rootstocks and in evolving new cultivars through hybridization. Litchi seeds show recalcitrant storage behavior and remain viable for four to six days at room temperature. Seed loses viability rapidly on drying and is killed by exposure to chilling temperatures.

Seed Collections

Litchi grows well in subtropical conditions, with high summer and low winter temperatures, and in the tropics, particularly at higher altitudes. Low humidity and dry winds cause fruit splitting. Litchi prefers deep, well-drained soil. Trees raised from air layering take four to six years to yield, as compared to about twenty years for those raised from seedlings. Self- and cross-pollination occur, and flies, wasps, ants, and bees cause pollination. Fruits take about two months to mature and change their color on maturity. Fruits are allowed to ripen on the tree. Large and healthy fruits are selected for seed purposes. Seed loses viability immediately after extraction; therefore, it is planted immediately or processed for short-term storage.

Seed Germination

Litchi seeds germinate immediately after extraction (Chandler, 1958; Bolt and Joubert, 1968; Higgins, 1971). Seed germination is poor and erratic (Menzel, 1985). Seedling emergence is relatively slow and nonuniform due to the hard seed coat. Seed germination is higher at constant temperatures of 25 to 30°C, and it takes 16 days to complete. Seed germination and seedling emergence is rapid in large seeds (Kadman and Slor, 1974). Seeds contain high amounts of starch, and its breakdown precedes germination. Starch degradation is a characteristic feature of germination. Amylase is active at the initiation of germination, and retention of high amounts of starch in cotyledons denotes its partial breakdown, resulting in poor mobilization to growing seedlings (Prasad et al., 1994).

Fruit maturity. Seeds develop rapidly immediately after fertilization, which is followed by the formation of the aril and membranous mesocarp. Seed germination improves with advance in maturity stage. Fruit color changes from yellow-green to dark reddish brown on maturity. Seed attains maximum dry weight eight weeks after anthesis and gives 42 percent germination. Highest seed germination (100 percent) was recorded in fruits during the ninth week after anthesis (Ray and Sharma, 1987).

Seed drying. Litchi seeds are susceptible to desiccation injury. They lose germination capacity and vigor potential on drying. The seed coat and cotyledon turn brown on drying after four to six days of storage at room temperature. Initial seed moisture content is 28.5 percent and viability is 100 percent immediately after seed extraction. Seed viability is lost completely if moisture drops to 19 percent. It appears that 20 percent moisture is critical for litchi seed storage; below this level seed viability is injured (Ray and Sharma, 1987). During this period, peroxidase content increases and ascorbic acid decreases (Chen and Fu, 1989).

Storage Conditions

Litchi seeds lose viability rapidly on detachment from fruit and are recalcitrant in storage behavior. They can be stored for a few weeks to a few months with prevailing storage methods. King and Roberts (1979) reported that such factors as drying injury, chilling damage, microbial contamination, and germination of high-moisture seeds contributed to loss of viability in litchi. Suitable storage methods need to be developed accordingly to meet these requirements in extending seed longevity.

Seed storage in fruit. Fruit storage is a simple method and is suitable for short-term maintenance of seed viability. The fruit's rough, thick skin slows loss of moisture and keeps the fruit in good condition. Ray and Sharma (1987) reported that seeds retained in fruit treated with benomyl (0.05 percent) and wax emulsion (6 percent) and sealed in polyethylene bags maintained 42 percent viability up to 24 days of storage.

Seed storage in water. Rapid loss of moisture is the main cause for poor storage of litchi seeds. Seeds stored in water gave 75 and 30 percent germination after 7 and 15 days of storage, respectively (Ray and Sharma, 1985).

Seed storage in gases. Singh and Prasad (1991) observed rapid loss of seed viability both at ambient temperatures and in the refrigerator. Further, storage in moisture-proof containers caused seed sprouting. Seeds stored in the open remained viable (15 percent) for five days (Ray and Sharma, 1985). Seeds stored in sealed polyethylene bags gave 50.7 percent germination after ten days (Ray and Sharma, 1987). Sowa, Roos, and Zee (1991) proposed a new approach to recalcitrant seed storage as in a combination of nitrous oxide (N_2O) and oxygen (O_2). However, seed storage in only anesthetic gas is not beneficial. When combined with oxygen (80 percent N_2O + 20 percent O_2) it resulted in 92 percent germination after three months of storage at 8 to 10°C, compared to 44 percent germination in seeds stored in air.

Seed storage with chemicals. Seed pelletization is useful in reducing moisture loss. Seeds soaked in calcium chloride (4 percent) followed by pelletization with sodium alginate (4 percent) preserved viability (53 percent) for 80 days at 10°C (Yu et al., 1993).

REFERENCES

Bolt, L.C. and Joubert, A.J. 1968. Litchi grafting trials. *Farming S. Afr.* 44:11-13.
Chandler, W.H. 1958. *Evergreen orchards.* Philadelphia: Lea and Febiger.
Chen, G.Y. and Fu, J.R. 1989. Deterioration of some recalcitrant seeds. *Pl. Physiol. Comm.* 3:11-14.
Higgins, J.E. 1971. Litchi in Hawaii. *Bull. Hawaii Agric. Exp. St.* 44:1-21.
Kadman, A. and Slor, E. 1974. Experiments with propagation of the litchi *(Litchi chinensis)* in Israel. *Indian J. Hort.* 31:28-33.
King, M.W. and Roberts, E.H. 1979. *The storage of recalcitrant seeds: Achievements and possible approaches.* Rome: International Board for Plant Genetic Resources, pp. 1-95.
Menzel, C.M. 1985. Propagation of lychee: A review. *Scientia Hort.* 25:31-48.
Prasad, J.S., Mishra, M., Kumar, R., Singh, A.K., and Prasad, U.S. 1994. Enzymic degradation of starch in germinating litchi *(Litchi chinensis)* seeds. *Seed Res.* 22:89-97.
Ray, P.K. and Sharma, S.B. 1985. Viability of *Litchi chinensis* seeds when stored in air and in water. *J. Agril. Sci.* 104:247-248.
Ray, P.K. and Sharma, S.B. 1987. Growth maturity, germination, and storage of litchi seeds. *Scientia Hort.* 33:213-221.
Singh, A.K. and Prasad, U.S. 1991. Dehydration pattern and viability loss in seeds of two cultivars of litchi *(Litchi chinensis). Seed Res.* 19:41-43.

Sowa, S., Roos, E.E., and Zee, F. 1991. Anaesthetic storage of recalcitrant seeds: Nitrous oxide prolongs longevity of lychee and longan. *HortSci.* 26:597-599.

Yu, X.P., Fang, J., Sheng, S.J., Sun, C.K., Xu, X.Y., and Zheng, J.B. 1993. Viability of *Litchi chinensis* seeds stored in sodium alginate pellets. *Proc. 4th Nat. Workshop Seeds*, Angers, France, pp. 20-24.

– 13 –

Loquat: *Eribotrya japonica* Lindl.

Introduction

Loquat is a popular dessert fruit in the subtropics. Fruit is eaten fresh and processed for jam, juice, squash, and jelly. Fruit contains high amounts of carbohydrates, phosphorus, calcium, iron, and vitamin A.

Origin and Distribution

Loquat originated in China. It is cultivated predominantly in Australia, eastern Asia, South Africa, South America, and the Mediterranean region. Some major loquat-producing countries are Australia, China, Italy, Japan, and the United States.

Morphology

Loquat belongs to family Rosaceae and has chromosome number $2n = 34$. It is an evergreen, dicotyledonous, dense tree that grows up to 7 m in height. Leaves are large, alternate, lanceolate to obviate, with a shiny upper surface. Flowers are white, fragrant, and borne on terminal panicles; the calyx is five lobed; petals number five; stamens number twenty; and pistil five joined toward the base. Fruits are borne in cluster; they are round, oval, or pyriform and yellow to orange in color. Flesh is deep orange and juicy. Seeds are large, smooth, brown, and oval; they normally number three to five in a fruit, rarely up to ten.

Seed Storage

Loquat is propagated by vegetative means, such as air layers and in-arch grafting. Seed germinates readily, but seedling trees bear inferior fruits that are not true to type. Seeds are used for raising rootstocks and for evolving new cultivars through hybridization. Loquat seeds show recalcitrant storage behavior and do not withstand loss of moisture and exposure to chilling temperatures during storage (King and Roberts, 1979).

Seed Collections

Loquat grows well in mild subtropical climates. Temperatures below the freezing point are harmful. Trees are resistant to heat and drought conditions, preferring well-drained soil. Rootstocks are commonly raised from seeds. Self-incompatibility occurs in many loquat cultivars and invites cross-pollination. Trees bear fruit from the third year onward. Fruits are formed in clusters and harvested when all are mature. Large, uniform, and healthy fruits are selected for seed purposes. Seeds are removed from fruits, washed, and prepared for storage.

Seed Germination

Seeds require 21 days for germination. Seedling emergence is rapid, uniform, and higher when seeds are exposed to chilling at 1 to 5°C for 30 to 60 days (Ellis, Hong, and Roberts, 1985).

Storage Conditions

Loquat seeds are short-lived under ambient conditions. Seeds deteriorate rapidly after one month of storage at 5°C. Seeds stored in moist conditions at 5°C and high relative humidity gave 92 percent germination after six months of storage (Zink and Ojima, 1965).

REFERENCES

Ellis, R.H., Hong, T.D., and Roberts, E.H. 1985. *Handbook of seed technology for gene bank,* Volume II. Rome: International Board for Plant Genetic Resources, p. 556.

King, M.W. and Roberts, E.H. 1979. *The storage of recalcitrant seeds: Achievements and possible approaches.* Rome: International Board for Plant Genetic Resources, pp. 1-95.

Zink, E. and Ojima, M. 1965. The effect of storage conditions on the germinating capacity of loquat seeds. *Bragantia* 24:9-12.

Longan: *Euphoria longan* Steud.

Introduction

Longan is also called cat's eyes. It is cultivated widely in Southeast Asia. Fruit is sweet and tasty and the juicy aril is eaten fresh. Fruit is also dried, canned, or frozen. The tea made from longan fruit is considered a tonic. Dried flowers are used for medicinal purposes, and the wood is a valuable timber for furniture. Seeds have a high saponin content and are used in manufacturing shampoo.

Origin and Distribution

Longan is native to China. It is cultivated in Australia, China, India, Indonesia, Java, Malaysia, the Philippines, Sri Lanka, Thailand, the United States, and Vietnam.

Morphology

Longan belongs to family Sapindaceae and has chromosome number $2n = 30$. Trees are evergreen, dense, and grow up to a height of 40 m. The inflorescence is terminal; flowers are unisexual, yellowish-brown, and self-incompatible. Fruits are globular and smooth and contain black seeds.

Seed Storage

Longan is propagated by seeds and grafting. Seeds remain viable for a short period under ambient conditions. Longan seeds are recalcitrant in storage behavior and lose viability on drying and preserving at chilling temperatures (Ellis, Hong, and Roberts, 1985).

Seed Collections

The longan tree grows well in cool climates and at an altitude up to 1,000 m. Seedlings and in-arch grafts are used for raising orchards. It is self-incompatible, and bees bring about pollination. Fruit requires five months from flowering to harvest. Ripe and healthy fruits are selected for seed purposes. Seeds are removed, washed, and used immediately for propagation or for short-term storage.

Seed Germination

Seed germination and seedling development is slow and nonuniform due to the hard seed coat. Seedling emergence is faster and higher at 25 to 30°C in presence of light. Germination takes about 17 days to complete (Ellis, Hong, and Roberts, 1985).

Storage Conditions

Longan seeds are short-lived and difficult to preserve for long periods. Chen and Fu (1989) reported that seeds remain viable for four to six days at room temperature. Seeds experience rapid moisture loss, a decrease in ascorbic acid, and an increase in peroxidase content.

Seed storage in fruit. Fruit storage is a simple method for seed storage. Fruits treated with benomyl (0.05 percent) stored in polyethylene bags retained viability for five days. Further, longevity is extended with refrigeration (Wong, 1992).

Seed storage with gases. Seed storage with a combination of nitrous oxide and oxygen preserved viability relatively longer. Sowa, Roos, and Zee (1991) reported that seeds stored in 80 percent N_2O + 20 percent O_2 remained viable (70 percent) for seven weeks, as compared to nil with air storage.

REFERENCES

Chen, G.Y. and Fu, J.R. 1989. Deterioration of some recalcitrant seeds. *Pl. Physiol. Comm.* No. 3:11-14

Ellis, R.H, Hong, T.D., and Roberts, E.H. 1985. *Handbook of seed technology for gene bank,* Volume II. Rome: International Board for Plant Genetic Resources, p. 653.

Sowa, S., Roos, E.E., and Zee, F. 1991. Anesthetic storage of recalcitrant seeds: Nitrous oxide prolongs longevity of lychee and longan. *HortSci.* 26:597-599.

Wong, K.C. 1992. Studies on the propagation, seed storage and fruit storage of *Dimocarpus longan. Acta Hort.* 292:69-71.

– 15 –

Mangosteen: *Garcinia mangostana* L.

Introduction

Mangosteen is a popular delicious fruit in Southeast Asia. Pulp is eaten fresh, canned, and made into squash, syrup, or jelly. It is also used for medicinal purposes. The wood is valued for high-quality timber in the preparation of a wide range of furniture.

Origin and Distribution

Mangosteen is native to Malaysia. The wild species of mangosteen are found in Malaysia and India. Its cultivation was initially done in peninsular Malaysia. Later it spread to other tropical countries, but with limited success (gave low yield and poor fruit quality). Mangosteen is grown in Cost Rica, Ecuador, Honduras, India, Indonesia, Kampuchea, Madagascar, Malaysia, Myanmar, Panama, the Philippines, Sri Lanka, Thailand, and Vietnam.

Morphology

Mangosteen belongs to family Guttiferae and has chromosome number $2n = 76$. An evergreen tree is glabrous, dioecious, and slow growing, reaching a height up to 15 m. Leaves are opposite and ovate. Flowers are unisexual, yellow, and borne terminally; sepals number four; petals number four, and the ovary is four to eight celled. Fruit is a subglobose berry with persistent calyx and stigma lobes; pericarp is purple and tough with a yellowish resin and a specific flavor. Seeds vary in number from none to three and are formed from nucellar tissue.

Seed Storage

Vegetative propagation is seldom successful in this crop; hence, seeds are used for raising plants. Seeds have low and short viability after removal from fruit. Mangosteen seeds show recalcitrant storage behavior. Seeds are killed on desiccation and by exposure to chilling temperatures.

Seed Collections

Mangosteen grows well in hot, humid climates where temperatures range from 25 to 35°C and the humidity is higher than 80 percent. Mangosteen adapts to high rainfall conditions. Trees raised from seedlings are true to type, as seeds are formed parthenocarpically from nucellar tissue. Plants bear fruit 10 to 15 years after planting. Healthy and ripe fruits are collected for seed purposes. Seeds are difficult to preserve, even for a week. They are therefore used immediately upon removal from fruit.

Seed Germination

Mangosteen seeds germinate readily and do not show any dormancy. Seed germination is rapid and higher at a constant temperature range of 25 to 35°C in the presence of continuous light, and it is completed in 28 days (Ellis, Hong, and Roberts, 1985).

Storage Conditions

Mangosteen seeds are recalcitrant in storage behavior and lose viability quickly when the thin membrane around the seed is damaged and subjected to drying or low storage temperatures (Chin and Roberts, 1980). Seed viability is lost by four weeks after air-drying and by one week in a desiccator (Winters and Rodriguez, 1953). Seeds stored in fruit show delayed germination. Mangosteen seeds along with fruit stored better at 25°C than at 15°C or 35°C. Seeds stored in a humid atmosphere at 25°C maintained viability (56 percent) for seven weeks. Moist or dry seeds did not germinate after one week of storage at 10°C. However, seeds stored in moistened charcoal at room temperature gave highest percentage of germination (Winters and Rodriguez, 1953). Cox (1976) observed that seed storage in moist charcoal or moss at room temperature preserves viability for a few months.

REFERENCES

Chin, H.F. and Roberts, E.H. 1980. *Recalcitrant crop seeds.* Kuala Lumpur, Malaysia: Tropical Press.

Cox, J.E.K. 1976. Garcinia mangostana—Mangosteen. *Propagation of tropical fruit trees,* In Garner, R.J. and Chaudhri, S.A. (Eds.). Slough, England: Commonwealth Agricultural Bureaux, pp. 361-375.

Ellis, R.H., Hong, T.D., and Roberts, E.H. 1985. *Handbook of seed technology for gene bank,* Volume II. Rome: International Board for Plant Genetic Resources, p. 653.

Winters, H.F. and Rodriguez, C.F. 1953. Storage of mangosteen seeds. *Proc. Am. Soc. Hort. Sci.* 61:304-306.

Passionfruit: *Passiflora edulis* Sims

Introduction

Passionfruit is widely cultivated in the tropics and subtropics in small orchards or in home gardens. The pulp and seeds are eaten fresh from shell and used in salad. The flavored pulp is used in the preparation of jams, jellies, and beverages. Seeds contain a high amount (20 percent) of edible oil.

Origin and Distribution

Passionfruit originated in South America, where wild types are abundant. The two common types of passionfruit are purple and yellow. Purple passionfruit is native to southern Brazil. The yellow type evolved from the purple one as a natural mutant. The purple cultivars grow better at higher altitudes in the tropics, while the yellow types grow better in lower areas. Passionfruit is cultivated in Argentina, Australia, Brazil, Mexico, New Zealand, South Africa, the United States, and the West Indies.

Morphology

Passionfruit belongs to family Passifloraceae and has diploid chromosome number 2n = 18. It is a woody, vigorous, dicotyledonous, perennial climber growing up to 15 m long. Leaves are dark green, palmate, and three lobed with axillary tendrils. Flowers are perfect, single, axillary, and fragrant with showy calyx; they are tubular in shape, with five petals and five stamens, and the ovary has single locule. Fruit is a berry, globose or ovoid, deep purple with many dots, hard pericarp. Seeds are numerous, flat, and surrounded by yellowish pulp and a juicy aril.

Seed Storage

Seeds are largely used for propagation of the plant. They are also used in evolving new cultivars through hybridization and in the conservation of genetic diversity. Seeds are also eaten as a food along with the pulp. Passionfruit seeds show orthodox storage behavior and tolerate loss of moisture and chilling during storage. However, seeds deteriorate rapidly under high humidity tropical conditions.

Seed Collections

Passionfruit grows well in hot humid climates and in a wide range of soils. Seeds are sown in a nursery. Seedlings are transplanted to the field after three to four months of sowing. Suitable supports, such as trellises or poles, are provided for optimum vine growth (see Figure 16.1). Plants flower on new growth, and bees cause pollination. Fruit normally takes ten weeks for development. Healthy and ripe fruits are selected for seed purposes. Seeds are separated from fruit, washed thoroughly, dried, cleaned, and packed for storage.

Seed Germination

Seed germination is slow and nonuniform. Fresh seeds show a considerable amount of dormancy due to the presence of a hard seed coat. Seed scarification using sand paper and removal of the hard seed covering promotes germination (Morley-Bunker, 1980). Further, percentage of germination improves with exposure to alternate temperatures of 20/30°C for 16/8 h for six weeks. In normal conditions, seeds take 12 weeks for seedling emergence.

FIGURE 16.1. Passionfruit Vines Grown on Trellises

Storage Conditions

Seed deteriorates rapidly at higher storage temperature and higher relative humidity. Costa, Oliveira, and Lellis (1974) reported that seeds stored at room temperature gave higher germination after 240 days of storage after drying seeds with pulp in sun or without pulp in shade. In open storage, seeds lose viability rapidly. Seeds with 5.2 percent mc stored at room temperature lost viability completely after ten months of storage (Teng, 1977). Low-moisture seeds packed in polyethylene bags and stored in cool conditions preserved high viability for longer periods. Seeds stored in plastic bags in an air-conditioned room maintained high viability and exhibited 72 percent germination after 12 months of storage. Oliveira, Sadar, and Zampieri (1984) observed loss of viability to 50 percent after five years of ambient storage in paper bags, and none of the seeds germinated during the sixth year. Seeds packed in polyethylene bags and stored in cool chambers gave 60 percent germination after 57 months of storage, unlike none after 32 months in paper bags in ambient conditions (Nakagawa, Cavariani, and Amaral, 1991).

REFERENCES

Costa, C.F.Da., Oliveira, E.L.P.G.De, and Lellis, W.T. 1974. Persistence of the germinating capacity of passionfruit seeds. *Boletim do Instituto Biologico da Bahia* 13:76-84.

Morley-Bunker, M.J.S. 1980. Seed coat dormancy in *Passiflora* species. *Ann. J. Royal New Zealand Inst. Hort.* 8:72-84.

Nakagawa, J., Cavariani, C., and Amaral, W.A.N.Do. 1991. Storage of passionfruit seeds. *Revista Brasileira de Sementes* 13:77-80.

Oliveira, J.C.De, Sadar, R. and Zampieri, R.A. 1984. Effect of age on the emergence and vigor of yellow passionfruit seeds. *Revista Brasileira de Sementes* 6:37-43.

Teng, Y.T. 1977. Storage of passionfruit *(Passiflora edulis forma flavicarpa)* seeds. *Malaysian Agril. J.* 51:118-123.

REFERENCES

17

Sugar Apple: *Annona squamosa* L.

Introduction

Sugar apple is also called custard apple or sweetsop. It is cultivated for the edible fruits. Ripe fruits are eaten fresh and used in the preparation of juice and ice cream. Sugar apple originated in South America and is now widely grown in Asia and Africa. It is presently cultivated in Brazil, Costa Rica, India, Indonesia, the Philippines, and Seychelles. A member of the family Annonaceae, it has chromosome number $2n = 14$. It is a small tree growing up to 5 m in height. Flowers are perfect and yellowish in color. Fruit is a fleshy syncarp formed by the fusion of pistils and receptacle that is yellowish green, heart shaped, and covered with fleshy tubercles. Pulp is sweet, whitish, and embedded with numerous hard seeds. Pulp contains 16 to 18 percent sugar. Seeds are commonly used for propagation, and seedling trees are heterogeneous. Flowers are protogynous, and pollination is by insects. Ripe fruits are soft and perishable.

Seed Storage

Sugar apple seeds are orthodox in storage behavior (Ellis, 1984) and preserve fairly well under ambient conditions. Seeds show a short period of dormancy. The embryo is small, partially developed, and embedded in a large endosperm. It continues to develop even after shedding from the mother plant, and one to three months are required for after-ripening of seeds. About 60 percent of seeds are viable after two months of storage (Hayat, 1963). Low-moisture seeds packed in moisture-proof containers at low temperatures maintain viability for longer periods.

REFERENCES

Ellis, R.H. 1984. Revised table of seed storage characteristics. *Plant Genetic Resources Newsletter* 58:16-33.
Hayat, M.A. 1963. Morphology of seed germination and seedling in *Annona squamosa. Bot. Gaz.* 124:360-362.

18

Soursop: *Annona muricata* L.

Introduction

Soursop is commonly cultivated for the edible fruits. Fruit is eaten fresh after peeling the exocarp. Flesh is used in preparation of juice. Soursop is native to South America and cultivated throughout the tropics. It is largely grown in Surinam, the United States, and Venezuela. Soursop is a small evergreen tree belonging to the family Annonaceae and has chromosome number 2n = 16. Flowers are yellow, protogynous, and strongly odorous. Fruit is dark green, ovoid, and covered with curved fleshy spines. Pulp is soft white, with a sweetish sour taste, and contains many black seeds. It is propagated by seeds and grows well in sandy loam soils. It is cross-pollinated by insects.

Seed Storage

Soursop seeds show orthodox storage behavior (Ellis, 1984). Seed loses viability rapidly at higher temperature and relative humidity. Seed viability is lost completely after 210 days of storage at ambient temperature or 30°C, while it can be preserved for 390 days when stored in cloth, paper, plastic, or glass containers at 5 or 20°C (Lopes, Almeida, and Assuncao, 1982).

REFERENCES

Ellis, R.H. 1984. Revised table of seed storage characteristics. *Plant Genetic Resources Newsletter* 58:16-33.

Lopes, J.G.V., Almeida, J.I.L.De., and Assuncao, M.V. 1982. Preservation of germinative power in seeds of soursop *(Annona muricata)* under different temperatures and types of packing. *Proc. Trop. Reg. Am. Soc. Hort. Sci.* 25:275-280.

19

Date Palm: *Phoenix dactylifera* L.

Introduction

Dates are a popular dessert fruit in North Africa. They are cultivated for dried date fruits and also grown as ornamental plants in the semiarid tropics. Fully ripe dried dates are eaten fresh. Fruit is high in sugar, calcium, potassium, and iron contents. Plants require high heat during growing and ripening. Date palm originated in Central Asia, particularly in Iraq and surrounding areas. It is cultivated predominantly in the tropics and subtropics of western Asia and North Africa. Major date-producing countries are Algeria, Bahrain, Egypt, Iran, Iraq, Libya, Morocco, Pakistan, Saudi Arabia, Sudan, Tunisia, and the United States.

Date palm belongs to family Palmae. The diploid chromosome number is 2n = 36. Tree is slender, tall (30 m), dioecious, and erect with ascending foliage. Leaves are narrow and pinnate. Flowers are small and yellowish brown. Fruit is reddish brown, oblong or cylindrical, 2 to 8 cm long with sweet edible pulp and a single pointed grooved seed. Tree is propagated by offshoots or seedlings. It is wind pollinated, and fruiting occurs five to six years after planting.

Seed Storage

Seed Germination

Seed germination is slow and delayed, owing to the hard seed surface. Soaking in concentrated sulfuric acid for 30 min hastens seed germination.

Storage Conditions

Date palm seeds survive for short periods under ambient conditions. Seeds exhibit orthodox storage behavior. High seed moisture and high storage temperature reduce the storage life of seeds. Seeds having 15 percent moisture stored in paper bags deteriorated after three months, and in polyethylene bags at 3 to 4°C, seeds kept well for seven months (Araujo and Barbosa, 1992). Fresh seeds show 91.8 percent germination after one month of storage, but this was reduced to 55.2 percent after nine months of storage (Hore and Sen, 1996). Nixon (1964) preserved date seeds for six years at an average temperature of 10°C in winter and 29.4°C in summer. Seed germination was 26 percent after 11 years, and after 15 years of storage, none of the seeds germinated.

REFERENCES

Araujo, E.F. and Barbosa, J.G. 1992. Influence of packing and storage environment on preservation of palm *(Phoenix loueiri)* seeds. *Revista Brasileiria de Sementes* 14:61-64.

Hore, J.K. and Sen, S.K. 1996. Viability of date palm seeds under different prestorage treatments. *Curr. Res.* 25:54-56.

Nixon, R.W. 1964. Viability of date seeds in relation to age. *Rep. 41st Ann. Date Gr. Inst. Coachella* pp. 3-4.

20

Strawberry:
Fragaria ×*ananassa* Duch.

Introduction

Strawberry is a cool-season and cool-region crop. It grows well in cooler mountainous areas. It is largely cultivated in North America and Europe. The major strawberry-growing countries are Australia, Canada, England, France, Germany, Israel, Italy, Japan, New Zealand, Russia, Turkey, and the United States. Ripe fruits are eaten fresh or served with ice cream or in fruit salad. Fruit is aggregate and rich in vitamin C and amino acids. Strawberry is a stoloniferous perennial herb belonging to family Rosaceae. The chromosome number is 2n = 16. Leaves are light green and large; flowers are yellow and are self- and cross-pollinated. Fruit is an aggregate of small achenes, sweet, and red to dark red. Seeds and runners are used for propagation. Seeds are extracted from ripe fruits, washed, dried, and packed for storage.

Seed Storage

Strawberry seeds show orthodox storage behavior. Seed longevity improves with reduction in moisture and storage at lower temperatures.

Seed Germination

Seed germination is slow and erratic, owing to dormancy (Henry, 1934). Fresh seeds of different varieties gave 9 to 99 percent germination. Dormancy varies in different cultivars. Seeds produced by self-pollination are much more dormant than seeds produced from cross-pollination. Seed germination is improved with dry storage for three months, but this did not eliminate the dormancy (Adam and Wilson, 1967). Brown and Musa (1980) reported that seeds extracted from rotted fruits were less dormant, and seed drying at room temperature (18 to 20°C) promoted dormancy. Precooling of seeds eliminates the dormancy. Seeds exposed to 4°C for four weeks (Adam and Wilson, 1967) and in combination with GA (100 ppm) (Kretschmer and Krugersteden, 1997) showed higher germination. Light is essential for promotion of germination in dormant seeds. Red light promotes, and far-red light inhibits, the germination process (Toole, 1961). Nakamura (1972) opined that white light is sufficient for practical purposes in enhancing seedling emergence. According to Negi and Singh (1973), inhibitors were of nonphenolic type and removed by continuous washing of seeds in water. Seed soaking in ethephon (5,000 ppm) for 24 h followed by chilling for 30 days resulted in higher seed germination (Iyer and Subramanyam, 1976). Iyer, Chacko, and Subramanyam (1970) reported that ethephon replaced the chilling requirement and gave higher germination. Further, seeds sown in intermittent mist gave early and uniform seedling emergence (Iyer, Subramanyam, and Singh, 1975). Seed germination is also improved with soaking seeds in nitric acid and hydrogen peroxide (Negi and Singh, 1972). Seed scarification (sulfuric acid for 10 min) followed by chilling (4 to 6°C for five months) improved the germination percentage (Yamakawa and Noguchi, 1994).

Storage Conditions

Strawberry seeds are stored for two years under ambient conditions. Seed viability was higher in seeds stored in cooler locations (Myers, 1954). Seed viability is retained for 17 years in cool storage. Seed germination was early and rapid in older seeds (Moore and Scott, 1964). Low-temperature storage increases the storage life of seeds. Seeds preserved high viability and vigor at 4.4°C for two to three years (Scott and Draper, 1970).

REFERENCES

Adam, J. and Wilson, D. 1967. Factors affecting the germination of strawberry seeds. *A.R. Long Ashtion Agric. Hort. Res. Stn,* pp. 90-95.

Brown, A.E. and Musa, M.J. 1980. The beneficial effect of rotting of strawberry fruit by *Botrytis cinerea* on subsequent germination. *Seed Sci. Technol.* 8:269-275.

Henry, E.M. 1934. The germination of strawberry seeds and the technique of handling the seedlings. *Proc. Am. Soc. Hort. Sci.* 31:431-433.

Iyer, C.P.A., Chacko, E.K., and Subramanyam, M.D. 1970. Ethrel for breaking dormancy of strawberry seeds. *Curr. Sci.* 39:271-272.

Iyer, C.P.A., Subramanyam, M.D., and Singh, R. 1975. Improving seed germination with mist. *Curr. Sci.* 44:895-896.

Iyer, C.P.A. and Subramanyam, M.D. 1976. Interaction of chilling and ethrel on strawberry seed germination. *3rd Inter. Symp. Trop. Sub-Trop. Hort.* 2:139-145.

Kretschmer, M. and Krugersteden, E. 1997. Germination capacity of strawberry seeds. *Gemuse* 33:417-418.

Moore, J.N. and Scott, D.H. 1964. Longevity of strawberry seeds in cool storage. *Proc. Am. Soc. Hort. Sci.* 85:341-343.

Myers, A. 1954. Viability of strawberry seeds. *Agr. Gaz. N.J. Wales* 65:31.

Nakamura, S. 1972. Germination of strawberry seeds. *J. Jap. Soc. Hort. Sci.* 41:367-375.

Negi, S.P. and Singh, R. 1972. Effect of different chemicals on germination of strawberry seeds. *Indian J. Hort.* 29:265-268.

Negi, S.P. and Singh, R. 1973. Preliminary studies on inhibitors in strawberry seeds. *Indian J. Hort.* 30:370-375.

Scott, D.H. and Draper, A.D. 1970. A further note on longevity of strawberry seeds in cold storage. *HortSci.* 5:439.

Toole, E.H. 1961. The effect of light and other variables on the control of seed germination. *Proc. Inter. Test. Seed Test. Assoc.* 26:659-673.

Yamakawa, O. and Noguchi, Y. 1994. Effect of storage conditions and seed production time on seed germination in strawberry. *Bull. Nat. Res. Inst. Veg. Orna. Pl. Tea Sr. A. Veg. Orna. Plants* 9:41-49.

– 21 –

Durian: *Durio zibethinus* L.

Introduction

Durian is a fruit crop largely cultivated in Southeast Asia, especially in Malaysia. It is cultivated for the edible fruits. The sweet arils are eaten fresh or used in the preparation of jam or fermented products. Seeds are eaten after boiling and roasting. Pulp is rich in carbohydrates and fat, and it quickly turns sour and rancid.

Durian originated in Southeast Asia, particularly in the Malaysian peninsula. It is cultivated predominantly in Indonesia, Kampuchea, Malaysia, the Philippines, Thailand, and Vietnam.

Morphology

Durian belongs to family Bombacaceae, and its chromosome number is 2n = 56. It is a large evergreen tree. Flowers are white or pink in color and arranged in cymes. Fruit is a large, spherical, ovoid, spiny, green to yellow capsule that splits into five valves. Seeds are large and covered with a sweet, pulpy, cream-colored, edible aril with one to six seeds per cell. Durian grows well in hot, humid climates with high rainfall. It prefers sandy loam soil. Seeds and vegetative means are used for propagation. It fruits seven years after planting. Moths and bats pollinate it. Fruit takes about three months to develop. It emits a bad odor similar to a rotten onion immediately after ripening.

Seed Storage

Durian seeds remain viable for a short period under ambient conditions. Fresh seeds germinate readily within three days of sowing. Durian seeds exhibit recalcitrant storage behavior (King and Roberts, 1979). They lose viability on drying and exposure to chilling temperatures. Fresh seeds contain about 44 percent moisture and lose viability when moisture falls below 21 percent. Further, a longer period of moist storage causes sprouting and

113

rotting of seeds (Teng, 1977). Seed storage in airtight containers at 20°C preserves viability (90 percent) for 32 days (Soepadmo and Eow, 1977).

REFERENCES

King, M.W. and Roberts, E.H. 1979. *The storage of recalcitrant seeds*. Rome: International Board for Plant Genetic Resources, pp. 1-96.
Soepadmo, E. and Eow, B.K. 1977. The reproductive biology of *Durio zibethinus*. *Garden's Bull.* 29:25-33.
Teng, Y.T. 1977. Effect of drying on the viability of rambutan and durian seeds. *Malaysian Agricultural Research and Development Institute Res. Bull.* 5:111-113.

22

Phalsa: *Grewia asiatica* Mast.

Introduction

Phalsa is a delicate, easily perishable fruit. Ripe fruits are eaten fresh and used for juice preparation. Phalsa belongs to family Tiliaceae and originated in India. It is cultivated in India, Malaysia, Pakistan, and Taiwan. Fruit is somewhat acidic and contains high amounts of vitamins A and C and minerals such as calcium, phosphorus, and iron.

Phalsa is a bushy deciduous tree growing up to 10 m in height. Flowers are perfect; sepals number four to five; petals number four to five; stamens number 70 to 80; there is one pistil. Fruit is round, drupe, red or purple in color, and contains one to two seeds. Seeds are used for propagation. Fruits are harvested when color changes to reddish brown. Seeds are extracted from ripe fruits, washed, dried, and suitably packed for storage.

Seed Storage

Seeds show orthodox storage behavior and maintain greater longevity under dry, cool conditions. Fresh seeds show dormancy, which terminates with after-ripening. Phalsa seeds are short-lived under ambient conditions. They lose viability rapidly at higher temperatures and humidity. Seed viability is unaffected by drying in sun or shade (Chacko and Singh, 1971). Verma and Rathore (1977) reported that seeds sown after four weeks of storage resulted in higher germination as compared to fresh seeds, whereas none of the seeds germinated after ten weeks of storage under ambient conditions. Seeds preserved higher viability when packed in sealed jars and stored at 5 to 8°C or room temperature than when stored in open conditions at the same temperature (Chacko and Singh, 1971). Well-dried seeds (8 percent mc) packed in polyethylene bags stored at 5 and –20°C maintained viability (60 percent) for ten years (Doijode, unpublished data).

REFERENCES

Chacko, E.K. and Singh, R.N. 1971. Studies on the longevity of papaya, phalsa, guava and mango seeds. *Proc. Inter. Seed Test. Assoc.* 36:147-158.

Verma, R. and Rathore, S.V.S. 1977. Note on seed storage studies in seed propagation of phalsa under Agra conditions. *Indian J. Agric. Res.* 11:185-187.

23

Macadamia Nut: *Macadamia integrifolia* Maiden & Betche

Introduction

Macadamia is cultivated for the edible seed kernels, which are eaten as a dessert or used in ice creams, chocolates, and confectioneries. It is rich in fat. Macadamia is native to Australia, and largely cultivated in Australia, Brazil, Costa Rica, Guatemala, Kenya, South Africa, the United States, and Zimbabwe.

Macadamia is an evergreen tree belonging to family Proteaceae. Flowers are small, perfect, and borne in large numbers (100 to 300). Fruit is a follicle, seeds are spherical with hard testae, and the kernel is creamy white and rich in fat. Macadamia grows well in high rainfall areas in the tropics. Many cultivars are self-incompatible and pollinated by honeybees. Fruits that fall on maturity are collected for storage.

Seed Storage

Seeds are used for propagation as well as for raising rootstocks. Macadamia seeds appear to be recalcitrant and viable for a short period under ambient conditions (King and Roberts, 1979). Seed viability varies from 42 to 83 percent in different cultivars. Fairly longer periods (40 to 240 days) are required for seedling emergence following sowing. Seeds are viable for four months at room temperature (Ojima, Dallorto, and Rigitano, 1976). Hamilton (1957) reported that seed viability was preserved for four months at room temperatures (6.7 to18.3°C), but it subsequently deteriorated and was lost completely by 12 months of storage. Similarly, moist seeds retained 50 percent of germination after six months of storage under ambient conditions (Storey, 1969).

REFERENCES

Hamilton, R.A. 1957. A study of germination and storage life of macadamia seeds. *Proc. Am. Soc. Hort. Sci.* 70:209-212.

King, M.W. and Roberts, E.H. 1979. *The storage of recalcitrant seeds.* Rome: International Board for Plant Genetic Resources, pp. 1-96.

Ojima, M., Dallorto, F.A.C., and Rigitano, O. 1976. Germination of macadamia seeds. *Bol. Tec. do. Inst. Agrono.* 33:16.

Storey, W.B. 1969. Macadamia. In *Handbook of North America nut trees,* Jaynes R.A. (Ed.). Geneva, NY: The W.F. Humphery Press, Inc., pp. 321-335.

24

Indian Jujube:
Ziziphus mauritiana Lam.

Introduction

Indian jujube is also known as ber. It is grown mainly for the edible fruits. The fruit is eaten fresh, dried, or canned. It is rich in minerals, such as iron, phosphorus, calcium, and in vitamins A and C. Leaves, bark, and roots are used for medicinal purposes. The wood is a valuable timber in making furniture. Indian jujube originated in India. Later it spread to Africa and Southeast Asia. It is commonly grown in China, India, Indonesia, Malaysia, Mozambique, Pakistan, the Philippines, Sri Lanka, and the West Indies.

Morphology

Indian jujube belongs to family Rhamnaceae and has chromosome number 2n = 48. It is a small, thorny evergreen tree. Leaves are simple, alternate, and broadly oval. Flowers are small, greenish, hermaphrodite, and borne in clusters at the leaf axils. Fruit is ellipsoid to subglobose drupe, fleshy, and orange to brown in color with edible acidic pulp and a hard central stone. The hard endocarp encloses one or two seeds.

Seed Storage

Seeds are predominantly used for propagation. They show orthodox storage behavior and maintain viability for fairly long periods under ambient conditions. However, unfavorable storage conditions reduce seed longevity. Seeds are also used for raising rootstocks for vegetative propagation by budding or grafting and in the conservation of genetic diversity.

Seed Collections

The Indian jujube grows well in hot, dry climates. High humidity and frost damage the crop, especially during the fruiting stage. It prefers deep, sandy loam soils. However, it withstands waterlogging and drought conditions. Plants are raised through seeds or patch budding. Honeybees cause pollination. Fruit takes four to five months to mature. Large healthy ripe fruits are used for seed purposes. Seeds are extracted, washed, dried, and suitably packed for storage.

Seed Germination

Seed germination is slow and uneven, owing to the hard endocarp. The seed is covered by the endocarp, which hardens after 50 days of bloom, and the testa turns to dark brown 100 to 110 days after anthesis. Moisture content decreases to 30 percent at harvest. Immature seeds do not germinate, while treatment with benzyladenine or gibberellic acid (10 ppm) stimulates germination (Kim and Kim, 1983). Seed germination is low in fresh seeds but improves on soaking in water or in acid. Seed soaking in concentrated sulfuric acid for six minutes gave highest seedling emergence (Singhrot and Makhija, 1979). Casini and Salvadori (1980) reported that removal of the endocarp and treatment with gibberellic acid (400 ppm) improved germination, and seed with the endocarp intact germinated after scarification by immersion in concentrated sulfuric acid.

Storage Conditions

Fresh seeds gave low germination (33 percent), even up to eight months of storage. Seedling vigor was high at the end of the storage period (Reddy and Murthy, 1990). Seed viability and vigor are retained with low-temperature storage. About 50 to 60 percent of seeds were viable after ten years of storage at 5 and –20°C (Doijode, unpublished data). Seed viability improved in one-year-old seeds by soaking them in water for 48 h followed by four days' stratification in moist sand (Kajal, 1983). Seed treatment with potassium dihydrogen phosphate (1 percent) improved germination in stored seeds (203 days after extraction) (Ghosh and Sen, 1988).

REFERENCES

Casini, E. and Salvadori, S. 1980. The germination of seeds of jujube *(Zizyphus sativa)*. I. Effect of gibberellic acid on seeds with and without endocarp. *Revista di Agricoltura Sub-Tropicale e Tropicale* 74:39-47.

Ghosh, S.N. and Sen, S.K. 1988. Effect of seed treatment on germination, seedling growth and longevity of ber *(Ziziphus mauritiana)* seeds. *South Indian Hort.* 36:260-261.

Kajal, R.S. 1983. Studies on the effect of sowing depth, seed and budding treatment on germination and budling growth in ber *(Zizyphus mauritiana)*. *Thesis Abstr. Haryana Agri. University* 9:176.

Kim, Y.S. and Kim, W.S. 1983. Studies on germination of *Zizyphus jujuba* seeds at different stages of growth and development. *South Korean Hort.* 25:47-53.

Reddy, Y.N. and Murthy, B.N.S. 1990. Studies on germinability and seedling vigor at different intervals of seed storage in ber *(Zizyphus mauritiana)*. *Indian J. Hort.* 47:314-317.

Singhrot, R.S. and Makhija, M. 1979. Vegetative propagation of ber *(Zizyphus mauritiana)*. III. Effect of time of sowing and acid treatment on ber seed germination and seedling performance. *Haryana J. Hort. Sci.* 8:168-172.

25

Mulberry (*Morus* spp.)

Introduction

Mulberry is cultivated for the edible fruits in many parts of the tropics and subtropics. It is cultivated in Canada, China, India, Japan, and the United States. It belongs to family Moraceae. The different species include *Morus alba, M. indica, M. latifolia, M. lhou, M. nigra,* and *M. rubra.* The black *M. nigra* is well suited for higher altitudes in the tropics. Ripe fruits are eaten fresh as a dessert and also cooked as a vegetable. Tree is monoecious or dioecious, small, and with thin, ovate, serrate leaves that are often deeply lobed. Fruit bunch is syncarpous with many drupes, 2.5 to 5.0 cm long, and white, red, or pink in color. Seeds or cuttings are used for plant propagation.

Seed Storage

Seedling emergence is slow and nonuniform due to the presence of dormancy. Two years of dry storage over calcium chloride ($CaCl_2$) removes the dormancy, especially in *Morus alba* (Takagi, 1939). Seed exposure to alternate temperatures of 20/30°C for 16/8 h promotes germination (Ellis, Hong, and Roberts, 1985). Further, seed irradiation with gamma rays (7.5 kilorad [kr]) improves the germination (Das, 1970).

Mulberry seeds show orthodox storage behavior and are short-lived under ambient conditions (Kasiviswanathan and Iyengar, 1970). High seed moisture is injurious to seed longevity, and viability is lost rapidly at room temperature (Gupta, 1988). Seeds stored in desiccators over $CaCl_2$ + 20 percent H_2O (water) or 43.8 percent H_2SO_4 (sulfuric acid) remained viable for 14 years as compared to 4 months in control. The relative humidity in desiccators was 48 percent, and seed moisture content was 4 percent (Takagi, 1939). Low-moisture seeds packed in moisture-proof containers at low temperatures retain viability for fairly long periods.

REFERENCES

Das, B.C. 1970. Effect of gamma radiation on germination and seedling development of mulberry. *Sci. Cult.* 36:60-61.

Ellis, R.H., Hong, T.D., and Roberts, E.H. 1985. *Handbook of seed technology for gene banks.* Rome: International Board for Plant Genetic Resources, Volume II, pp. 507-508.

Gupta, P. 1988. Viability of mulberry seeds during storage. *Seed Res.* 16:248-250.

Kasiviswanathan, K. and Iyengar, M.N.S. 1970. Effect of maturity of fruit, month of collection and storage on the viability of three varieties of mulberry seeds. *Indian J. Seric.* 9:31-37.

Takagi, I. 1939. On the storage of mulberry seeds. *Res. Bull. Tokyo Imp. Sericult.* 2:1-22.

26

Sapodilla:
Manilkara achras (Mill.) Fosberg

Introduction

Sapodilla is an important tropical fruit cultivated for its edible sweet pulp. Normally, the ripe fruits, which are rich in sugars, calcium, phosphorus, and iron, are eaten fresh. Unripe fruits are sources for extraction of chicle gum. Sapodilla belongs to family Sapotaceae and is native to Mexico. It is cultivated in Brazil, Cuba, India, Malaysia, Mexico, the Philippines, Sri Lanka, Surinam, Thailand, the United States, Venezuela, and the West Indies.

Sapodilla is a hardy evergreen tree requiring less attention for its cultivation. It grows well in hot humid climates, especially in tropical lowlands. Fruit is globose to ovoid and rusty brown; the yellowish brown flesh contains nil to 12 black seeds. Seeds are hard and easily separable from the pulp.

Seeds, layers, and grafts are used for raising sapodilla plants. Sapodilla seedlings grow slowly; hence, it is replaced with *Manilkara hexandra* as a rootstock.

Seed Storage

Sapodilla seeds have limited use. They are used for raising rootstocks, and in hybridization programs. Seeds are viable for a short period. It appears that sapodilla seeds are recalcitrant in storage behavior (Ellis, 1984). They lose viability rapidly on removal from fruit. Seedling emergence is rather slow and uneven, possibly due to the hard seed coat. Seed germination varies in different cultivars. It was highest in cv. Periyakulam (PKM) (58 percent) and least in cv. Dwarapudi (26 percent) (Ponnuswami, Irulappan, and Vijay, 1988). Seed germination was improved by application of gibberellic acid (50 ppm), indoleacetic acid (50 ppm), and thiourea (1 percent) (Farooqui, Nalwadi, and Sulladmath, 1971; Srivastava, 1984).

REFERENCES

Ellis, R.H. 1984. Revised table of seed storage characteristics. *Plant Genetic Resources Newsletter* 58:16-33.

Farooqui, A.A., Nalwadi, U.G., and Sulladmath, U.V. 1971. Effect of growth regulators on the germination of sapota *(Achras sapota)* seeds. *Mysore J. Agric. Sci.* 5:341-343.

Ponnuswami, V., Irulappan, I., and Vijay, K. A. 1988. A note on germination of seeds of sapota varieties *(Achras zapota)*. *South Indian Hort.* 36:193-194.

Srivastava, H.C. 1984. Germination of sapodilla *(Manilkara achras)* seeds. *Seed Res.* 12:122-124.

27

Rambutan: *Nephelium lappaceum* L.

Introduction

Rambutan is largely grown in Southeast Asia. Fruits have a sweet, acidic, edible aril, which is eaten fresh, canned, or mixed with other fruits. Seeds, which are eaten after roasting, are rich in fat and are hence used in manufacturing soap and candles. Different plant parts are used for medicinal purposes. Rambutan is native to Malaysia. It is cultivated in Australia, Honduras, Indonesia, Malaysia, Mexico, the Philippines, and Thailand.

Rambutan belongs to family Sapindaceae and has chromosome number 2n = 22. It grows well in hot, humid tropical conditions with high rainfall. Its cultivation is successful in its native area. It prefers well-drained soil. Trees are evergreen, monoecious or dioecious, and grow up to 20 m in height. Flowers are unisexual or perfect, small, greenish white, and arranged in panicles. Fruit is globose or ovoid with hairy prickles. Its flesh is juicy, white, sweet, or sourish. Seeds are globose and 2.5 to 3.5 cm in diameter. Rambutan is propagated by seeds and by bud grafting. Seedling trees take longer to bear fruit, five to six years, and are not to true to type. Insects such as bees and flies pollinate the flowers. Fruits are nonclimatric and harvested when fully ripe. Seeds are removed from fruit, washed, and used immediately for propagation or for short-term storage.

Seed Storage

Rambutan seeds are used for propagation as well as for raising rootstocks. Seeds are recalcitrant and do not store well in ambient conditions (Ellis, 1984). Fresh seeds contain a high amount of moisture (32.5 percent) and germinate readily (100 percent). Seedling emergence was rapid when placed horizontally. Seeds lose viability within one week of storage under ambient conditions due to drying. During ambient storage the moisture decreases to 13 percent after two weeks and seeds are killed (Teng, 1977). Seed germination is affected in the presence of a fruit arilloid. The juice of the arilloid also inhibits germination. However, seeds are preserved well in moist sawdust or charcoal for three to four weeks. They sprout readily during storage. Unwashed seeds or seeds treated with juice from the arilloid remained viable for four weeks in moist sawdust without sprouting (Chin, 1975).

REFERENCES

Chin, H.F. 1975. Germination and storage of rambutan *(Nephelium lappaceum)* seeds. *Malaysian Agril. Res.* 4:173-180.
Ellis, R.H. 1984. Revised table of seed storage characteristics. *Plant Genetic Resources Newsletter* 58:16-33.
Teng, Y.T. 1977. Effect of drying on the viability of rambutan and durian seeds. *Malaysian Agricultural Research and Development Institute Res. Bull.* 5:111-113.

TEMPERATE FRUITS

– 28 –

Apple: *Malus domestica* Borkh. and Pear (*Pyrus* spp.)

APPLE

Introduction

Apple is the leading temperate fruit crop of the world. It is primarily cultivated for the edible fruits. Fruits are eaten fresh or processed as jam, juice, and cider. Fruits are rich in sugar and also contain high amounts of minerals, such as calcium, phosphorus, iron, and potassium, and vitamin thiamine. Trees have showy flowers and are used in landscaping and in home gardens for ornamental purposes.

Origin and Distribution

Apple is largely cultivated in Europe and North America. It is likely derived from *Malus pumila* in southwestern Asia. The major apple-growing countries are Australia, Brazil, Canada, China, Denmark, Egypt, England, France, India, Israel, Italy, Japan, Mexico, New Zealand, Russia, and the United States.

Morphology

Apple belongs to family Rosaceae, and its somatic chromosome number is 2n = 34. Trees are of medium size; leaves are soft and less pubescent; flowers are small, borne in clusters, and bright pink in color. Fruit is large and sweet and contains brown seeds. A thousand seeds weigh 25 to 50 g.

Seed Storage

Apple is propagated by both sexual and asexual methods. Seeds are predominantly used for raising rootstocks but are also involved in evolving better varieties through hybridization. Seeds survive for fairly long periods under ambient conditions. They exhibit orthodox storage behavior. Fresh seeds do not germinate readily due to innate dormancy, and they require cold stratification for removing dormancy. Dormancy is helpful in extending storage life of seeds for short periods. Normally, apple fruits are stored for several months in cold storage, which helps to preserve seed quality.

Seed Collections

Apple is a cool-season crop and is widely grown in cooler regions, particularly at higher latitudes and altitudes. It requires chilling temperatures for induction of flowering. It prefers deep, fertile, and well-drained soil. Apple is commonly propagated by grafting using suitable dwarfing rootstocks. Many apple cultivars are self-incompatible and need suitable pollenizers. Honeybees largely bring about pollination. The grafted plant bears fruit in two to four years after planting. Ripe, healthy, good-sized fruits are collected for seed purposes. Seeds are removed from the fruit, thoroughly washed, dried at low temperature and relative humidity, and suitably packed for storage.

Seed storage in fruit. Apple seeds preserve viability and vigor in fruits up to 180 days. Fruits stored for shorter periods showed high seed quality. The longer the seeds remained in the fruit (150 to180 days), the higher their susceptibility to mildew. Seeds stored in fruit for 30 days showed maximum seedling growth and development (Kopan, 1967). Seedling emergence in seeds extracted from fruits kept in cold storage was higher than for those kept at room temperature (CampoDall'orto et al., 1978). It is advantageous to preserve fruits at 5°C for a short period prior to seed extraction (Ellis, 1982).

Seed Germination

Apple seeds exhibit dormancy and do not germinate immediately after extraction. Storing fruits at low temperature eliminates the dormancy, but it increases time to germination (Wills and Scriven, 1983). Seed dormancy can be overcome by exposing moist seeds to low temperature (5°C) for various periods (Karnatz, 1969). According to Abbot (1955), germination of dormant seeds is higher at lower temperatures (5 to 10°C). Radicles emerge at lower temperatures (Sanada, Yoshida, and Haniuda, 1980a) and hypocotyls develop at higher temperatures. Seeds of late-harvested fruit (Sanada, Yoshida, and Haniuda, 1980b) and rotten fruits need shorter periods of stratification (Pandey and Tripathi, 1977; Pustovoitova and Rudnicki, 1977).

Seed coat removal facilitates germination but produces more abnormal seedlings. Normal growth is resumed after a certain period or through stratification of seeds or seedling exposure to low temperatures (Flemion, 1934). The longer period of stratification is reduced by growing excised embryo or treating seed with ethephon (250 ppm) followed by 60 days exposure to low temperature, which gave high seed germination (Sinha, Pal, and Awasthi, 1977).

Early and rapid techniques for determination of viability in apple seeds have been developed. Seeds are soaked for 2 to 3 h in sulfuric acid, followed by soaking in benzyladenine (10 ppm) or equal volumes of benzyladenine (15 ppm) and gibberellic acid (150 ppm). Seeds germinate in the presence of continuous light at room temperature. The viable embryo germinates in three to four days of sowing, while nonviable ones deteriorate (Badizadegan and Carlson, 1966).

Storage Conditions

Temperature. Apple seed exhibits orthodox storage behavior. Higher seed moisture and temperature reduce the storage life of seeds. The estimated storage life of apple seeds with 5 percent mc is 100 years and 37 years at −18 and 5°C, respectively (Dickie, 1986). Seed moisture is lowered rapidly in apple seeds by exposing them to low relative humidity. It takes about eight days to reduce moisture from 50 (fresh seeds) to 5 percent (Omura, Sato, and Seike, 1978). Seed germination decreases with increase in moisture during storage. Seeds stored in a damp atmosphere show low germination, while seeds stored at 50 to 55 percent RH exhibit higher germination (Solovjeva, 1950).

Grzeskowiak, Miara, and Suszka (1983) recommended sealed storage of apple seeds at −3°C for longer retention of viability and vigor. According to Kaminski (1974), drying stratified seeds at a higher temperature (25°C) induces secondary dormancy, not the lower temperature (5°C). Therefore, Conner (1947) recommends storing seeds in moist sand at 0°C and not allowing the seeds to dry. Cromarty, Ellis, and Roberts (1982) suggested drying seeds at 15°C with 10 percent RH for storage. Seeds (8 to 10 percent mc) were stored at 10°C in jute or polyethylene bags up to 25 months. In jute bags, viability was reduced from 90 to 30 percent within nine months, and in polyethylene bags, seeds showed 90 to 95 percent germination after 25 months of storage (Karaseva, Karpov, and Monakhova, 1981). Seed viability can be preserved for two years in desiccators with calcium chloride at 2 to 10°C (Solovjeva and Kocjubinskaja, 1955). Seed longevity is extended further to seven years by packaging seeds in hermetically sealed bottles and

storing them at –3 to –5°C (Solovjeva, 1966). Seed drying to 5 percent
moisture and storage at –18°C are beneficial for longer retention of high
seed viability and vigor (Ellis, 1982).

Invigoration of Stored Seeds

Seedling emergence is slow and less with an increase in storage period.
Seed stratification in moist sand at 5°C followed by seed soaking in
gibberellic acid (100 milligrams per liter [mg·liter $^{-1}$]) for 12 h increased the
seed viability in three-year-old apple seeds (Wanic, Kawecki, and Naglica,
1968).

PEAR

Introduction

Pear ranks second after apple as a popular cultivated fruit in temperate
regions. The two important cultivated are *Pyrus* species *P. communis* and
P. pyrifolia. Pear cultivation is also extended to subtropical regions due to
its hardy nature and wider adaption. Fruits are used for table purposes and
processed for preparation of jam, jelly, juice, wine, and candied and sweet
pickles. Fruits are rich in organic acids, such as malic and citric acids; minerals, such as copper, sulfur, and phosphorus; and vitamins, such as thiamine, riboflavin, and vitamin C. Pear flowers are beautiful and the trees are
planted in house gardens for ornamental purposes.

Origin and Distribution

Wide genetic diversity exists among pear cultivars in Europe and Asia.
Pear originated in eastern Asia, probably in western China. European cultivars
are descendants of *P. communis,* while Chinese and Japanese cultivars are
derived from *P. pyrifolia.* Pear cultivation is mainly confined to temperate
and subtropical regions of the northern hemisphere. It is grown in Argentina, Australia, China, France, Germany, Italy, Korea, Romania, Russia,
South Africa, Spain, Switzerland, Turkey, and the United States.

Morphology

Pear belongs to family Rosaceae, and its somatic chromosome number is
2n = 34. The commonly cultivated varieties belong to *P. communis* and
P. pyrifolia, while *P. serotina* and *P. pashia* are used as rootstocks. Pear is a

small deciduous tree. Leaves are small, ovate, and margin serrated; flowers are showy, hermaphrodite, and white. Fruit is a pome, is pyriform and fleshy, and contains a stone. Seeds are formed inside the stones.

Seed Storage

Pear is commonly propagated by vegetative means such as budding and grafting. Seeds are primarily used for raising rootstocks. Seeds are also used in evolving better cultivars through hybridization and for conservation of genetic diversity in the gene bank. Pear seeds exhibit orthodox storage behavior, and their longevity is extended by lowering seed moisture and storing at low temperatures.

Seed Collections

Pear plants require mild winters and summers for optimum growth. They grow well in deep, well-drained, sandy loam soil. Trees bear fruit five to six years after planting. Many pear cultivars are self-incompatible and need suitable pollenizers. Bees bring about pollination. Fully matured, healthy fruits are collected for seed purposes. Seeds are easily extracted after softening the fruit. They are washed thoroughly, dried, cleaned, and suitably packed for storage or stratified for immediate sowing to raise seedlings for rootstocks.

Seed Germination

Fresh pear seeds do not germinate readily and need longer stratification to stimulate the germination process (Westwood and Bjornstad, 1968). Seed dormancy is removed by storing moist seeds at low temperature. Seeds stored at 5°C for 150 days showed high germination in many pear cultivars. The inhibitor contents decrease and auxinlike substances increase in seed coat and embryo during stratification (Shawky et al., 1978). The duration of stratification varies from 15 to 180 days in different species. The stratification period also depends on the location from which fruits are collected. Seeds from warm winter climates require less chilling than those from colder climates (Westwood and Bjornstad, 1968). Seed germination is high after removal of the seed coat because this causes the excised embryo to germinate quickly (Omura, Sato, and Seike, 1978). The seedlings initially show abnormal growth if seeds have not been stratified. Shen and Mullins (1983) reported that moist seeds placed at 4°C for four weeks, and subsequently at the soil surface in the presence of dim light, gave higher germination. Seed germination is also improved with seed soaking in gibberellic acid (150 ppm) and thiourea (5,000 ppm), followed by 28 days of stratification (Dhillon and Sharma, 1978).

Storage Conditions

Pear seeds readily absorb moisture from the atmosphere. High seed moisture reduces storage life of seeds. It takes eight days for seed moisture to decrease from 50 (fresh seeds) to 5 percent (Omura, Sato, and Seike, 1978). Low moisture (5 to 6 percent) is ideal for seed storage. The drying of stratified seeds results in low germination due to the induction of secondary dormancy. Seeds germinate fully when stratified (Westwood and Bjornstad, 1968). Pear seeds are dried at low temperature (15°C) and low relative humidity (10 percent RH) for long-term storage (Cromarty, Ellis, and Roberts, 1982). Seed viability is preserved for two years when seeds are packed in sealed bottles or stored in desiccators over calcium chloride at 2 to 10°C (Solovjeva and Kocjubinskaja, 1955). Seed longevity was extended further to seven years by preserving seeds at –3 to –5°C (Solovjeva, 1966). Seeds preserved in liquid nitrogen (–196°C) maintain viability for three years (Omura, Sato, and Seike, 1978). Seed viability was successfully preserved for 22 years at –5°C (Solovjeva, 1978). For better and longer storage of pear seeds, Ellis (1982) suggested storing fruits initially at 5°C for a short period prior to seed extraction, then reducing moisture content to 5 percent before packing in suitable moisture-proof containers and storing at subzero temperature (–18°C).

REFERENCES

Abbot, D.L. 1955. Temperature and the dormancy of apple seeds. *Proc. XIV Inter. Hort. Congr. Scheveningen* 1:746-753.

Badizadegan, M. and Carlson, R.F. 1966. A new viability test for apple seeds. *Hort. Rep. Mich. St. University* No. 30:10.

CampoDall'orto, F.A., Ojima, M., Rigitano, O., Scaranari, H.J., and Martins, F.P. 1978. Germination of apple seeds. *Bragantia* 37:83-91.

Conner, E.C. 1947. The storage and germination of apple seeds. *Agr. Gaz. N.S. Wales* 58:414-416.

Cromarty, A.S., Ellis, R.H., and Roberts, E.H. 1982. *The design of seed storage facilities for genetic conservation.* Rome: International Board for Plant Genetic Resources, p. 96.

Dhillon, B.S. and Sharma, M.R. 1978. Note on the effect of growth regulators on the germination of wild pear seeds. *Indian J. Agril. Sci.* 48:370-372.

Dickie, J.B. 1986. A note on the long-term storage of apple seeds. *Plant Genetic Resources Newsletter* No. 65:13-15.

Ellis, R.H. 1982. Seed storage and germination of apple and pear. *Plant Genetic Resources Newsletter* No. 50:53-61.

Flemion, F. 1934. Dwarf seedlings from non-after-ripened embryos of peach, apple and hawthorn. *Boyce Thompson Inst. Contrib.* 6:205-209.

Grzeskowiak, H., Miara, B., and Suszka, B. 1983. Long term storage of seeds of Rosaceae species used as rootstocks for cherry, plum, apple and pear cultivars. *Arboretum Kornickie* 28:283-320.

Kaminski, W. 1974. Secondary dormancy of apple seeds. Part II. The effect of temporary decrease of water content. *Prace. Instytutu Sandownicta w Skierniewicach.* 18:17-24.

Karaseva, L.G., Karpov, B.A., and Monakhova, Yu.V. 1981. Commercial storage of apple seeds. *Sadovodstvo* No. 3:19-20.

Karnatz, A. 1969. Storage and stratification of small quantities of apple seeds of valuable breeding material. *Z. Pflanzenzucht.* 62:138-144.

Kopan, V.P. 1967. Should seeds be stored in the fruit? *Sadovodstvo* No. 1:22.

Omura, M., Sato, Y., and Seike, K. 1978. Long-term preservation of Japanese pear seeds under extra low temperature. In *Long-term preservation of favorable germplasm in arboreal crops,* Akihama, T. and Nakajama, K. (Eds.). Fujimoto, Japan, pp. 26-30.

Pandey, D. and Tripathi, S.P. 1977. A note on the effect of the conditions of the fruits on the germination of apple seeds and growth behavior of the seedlings. *Prog. Hort.* 9:75-76.

Pustovoitova, T. and Rudnicki, R.M. 1977. Dormancy of apple seeds as related to the size of apples and the stage of their maturation. *Fruit Sci. Reports* 4:1-4.

Sanada, T., Yoshida, Y., and Haniuda, T. 1980a. Studies on the method of seed storage in apple breeding. I. Suitable method for short-term storage. *Bull. Fruit Tree Res. Stn. Morioka* 7:1-14.

Sanada, T., Yoshida, Y., and Haniuda, T. 1980b. Studies on the method of seed storage in apple breeding. II. Effective method for breeding dormancy of dry stored seeds. *Bull. Fruit Tree Res. Stn. Morioka* 7:15-31.

Shawky, I., Tomi, A.El., Rawash, M.A., and Makarem, A. 1978. Preliminary studies on the germination of *Pyrus communis* seeds. *Res. Bull. Aim Shams University* No. 826:12.

Shen, X.S. and Mullins, M.G. 1983. Seed germination in pear rootstock *Pyrus calleyana. Aust. Hort.* 81:50-51.

Sinha, M.M., Pal, R.S., and Awasthi, D.N. 1977. Effect of stratification and plant growth regulating substances on seed germination and seedling growth in apples. *Prog. Hort.* 9:27-30.

Solovjeva, M.A. 1950. Fruit seed storage. *Orchard and Garden* No. 10:24-28.

Solovjeva, M.A. 1966. Long term storage of fruit seeds. *Proc. 17th Inter Hort. Congr. Md.* 1:258-266.

Solovjeva, M.A. 1978. Method of storing fruit seeds and determining their storage quality. *Byulleten Vsesoy Ord. Len. Ins. Rasten. N.I. Vavilova* 77:64-65.

Solovjeva, M.A. and Kocjubinskaja, V.N. 1955. Effect of storage conditions of seeds on germination and yield of standard rootstocks. *Sad i Ogorod* No. 9:53-55.

Wanic, D., Kawecki, Z., and Naglica, I. 1968. The effect of GA on germination of 3 year old stratified seeds of the apple variety common Antonovaka. *Zeiz nauk Wyzoz Szleroln Olsztyn Ser A* No. 2, 24:171-177.

Wills, R.B.H. and Scriven, F.M. 1983. Relation between germination of apple seeds and susceptibility of fruit to storage breakdown. *J. Hort. Sci.* 58:191-195.

Westwood, M.N. and Bjornstad, H.O. 1968. Chilling requirement of dormant seeds of 14 pear species as related to their climatic adaptation. *Proc. Am. Soc. Hort. Sci.* 92:141-149.

Peach: *Prunus persica* (L.) Batsch, Plum (*Prunus* spp.), and Cherry (*Prunus* spp.)

PEACH

Introduction

Peach is a delicious fruit grown in temperate regions. The edible juicy fruits are eaten fresh and processed for squash preparations. The short stature of the plant and its showy flowers are valued for ornamental purposes, and it is thus planted in home gardens for landscaping. The wide genetic diversity of peach, including wild types, is found in China, from where it is assumed to have originated. Peach is predominantly cultivated in Argentina, Australia, Chile, China, England, France, Japan, Mexico, New Zealand, South Africa, Spain, Turkey, and the United States.

Morphology

Peach belongs to family Rosaceae and has diploid chromosome number 2n = 16. Trees are small; leaves are long, simple, deciduous, and alternate. Flowers are perfect and pink to red in color. Fruit is a spherical, fleshy drupe with juicy outer layer surrounding hard stones that contain seeds. Trees are short-lived, lasting eight to ten years, and they commence fruiting from the third year onward.

Budding is followed for propagation of peach plants, and seeds are used for raising rootstocks. Peach is a cross-pollinated crop, and insects bring about pollination. Fruit changes color on maturity. Seeds are separated from matured fruits, washed thoroughly in water, dried, and suitably packed for storage.

Seed Storage

Seed Germination

Seed germination and seedling emergence are higher in mature seeds. Deol, Chopra, and Grewal (1993) reported that seeds collected from fruits harvested 10 to 12 days before commercial maturity had significantly higher germination than the seeds of fruits collected at commercial maturity.

Peach seeds show dormancy and need suitable dormancy-breaking treatments for higher germination. The hard endocarp delays seedling emergence, and removal of the endocarp results in early germination. Prechilling of seeds is beneficial for promoting seed germination. Exposure of moist seeds to 3 to 5°C for three to four months eliminates dormancy. Warm stratification for a short period followed by cold stratification promotes germination in dormant seeds. Suszka (1967) reported that preserving the seeds for 14 days at 20°C followed by prechilling at 3°C resulted in higher germination.

Storage Conditions

Peach seeds show orthodox storage behavior. Reducing seed moisture and storage at low temperatures help maintain seed viability. Cromarty, Ellis, and Roberts (1982) recommended that seed moisture could be effectively reduced at a low temperature (15°C) and seed stored at low relative humidity (10 percent RH) without damaging seed quality. Seeds are successfully dried to 8 percent mc (Toit, Jacobs, and Strydom, 1979). Seeds with intact endocarp take longer to dry; hence, the endocarp is halved carefully before drying. Earlier, seeds were stored under moist conditions for short storage periods. Seeds with 8 to 12 percent mc were preserved at 3°C for more than eight weeks (Suszka, 1964). Ellis and Hong (1985) recommended storage of dry seeds in moisture-proof containers at − 20°C for longer storage life of seeds.

PLUM

Introduction

Plum is a delicious juicy fruit grown in temperate regions. Fruits are used for table purposes, canned, dried, and processed for jam, jelly, squash, and confectionery. Plums are a rich source of sugar and contain high quantities of vitamins A, thiamine, and riboflavin and minerals such as calcium, iron, and phosphorus. Trees are planted in home gardens for ornamental purposes. The different types of cultivated plums include American plum *(Pru-*

nus americana), cherry plum *(P. cerasifera),* damson plum *(P. insititia),* European plum *(P. domestica),* and Japanese plum *(P. salicina).* Plum originated in North America, China, and Europe. The major plum-growing countries are Austria, Bulgaria, China, France, Germany, Hungary, Italy, Japan, Mexico, Romania, Russia, Spain, Turkey, and the United States.

Morphology

Plum belongs to family Rosaceae and has somatic chromosome number 2n = 16. Tree is small; leaves are ovate; flowers are white; fruit is oblong to ovoid and variously colored. Most plum cultivars are self-incompatible and require pollenizers. Bees bring about pollination. Ripe, healthy, and firm fruits are selected for seed purposes. Seeds are removed from pulp, washed thoroughly with water, dried, and packed in suitable containers for storage. Propagation is commonly achieved by budding; seeds are used in raising rootstocks.

Seed Storage

Seed Germination

Seed germination is slow and it takes longer for seedlings to emerge. Fresh seeds show dormancy. Germination is improved by removal of the seed coat. Prechilling of imbibed seeds also helps to break seed dormancy. Seed exposure to high temperature (20°C) for 14 days followed by chilling at 3 to 5°C hastens the germination process (Suszka, 1967).

Storage Conditions

Plum seeds show orthodox storage behavior. Dry seeds preserve high viability for long periods at cooler conditions. Giersbach and Crocker (1932) preserved seeds under moist conditions at 7 to 10°C. Seed germination was 98 percent after 26 months and 16 percent after 53 months of storage. Ellis and Hong (1985) recommended seed drying at 15°C and 10 percent RH to 4 to 6 percent mc. Dried seeds are packed in moisture-proof containers and stored at −20°C for long-term seed conservation.

CHERRY

Introduction

Cherry is commonly cultivated in temperate regions. Primarily two types of cherries are grown, namely, sweet cherry *(Prunus avium)* and sour cherry *(P. cerasus).* Sweet cherry is mainly used for table purposes, but also in salads and juice preparations and for canning. Sour cherry is largely used for processing purposes, such as wine or juice making, canning, and freezing; it is also sun dried. Fruit is rich in sugar, calcium, potassium, magnesium, iron, zinc, and vitamins A and C.

Morphology

Cherry has its origin in southeast Europe. It belongs to family Rosaceae, and its diploid chromosome number is 2n = 16, 32. Trees are tall; flowers are white and perfect. Fruits are globular or oblong and red, with a fleshy outer layer surrounding the hard stones that contain the seeds. Cherry is propagated by budding and grafting. Seeds are used for raising rootstocks. Many cherry cultivars are self-incompatible and require pollenizers. The compatible cultivars are grown side by side. Honeybees bring about pollination. Fruits are harvested when they change color. Seeds are extracted from fully ripe fruits, washed thoroughly with water, dried, and packed for storage.

Seed Storage

Seed Germination

Fresh cherry seeds do not germinate due to the presence of innate dormancy (Michalska, 1982). Seedling emergence is delayed due to the presence of a hard endocarp, and it improves with removal of the endocarp. Seed exposure to prechilling at 3 to 5°C hastens the germination process (Suszka, 1967). The percentage of germination was higher at alternate temperature regimes of 3/5°C or 5/20°C for 16 and 8 h, respectively, than at a constant temperature of 3 or 5°C (Suszka, 1967). Warm stratification before chilling is beneficial for promoting higher seed germination. Exposure of cherry seeds 14 days at 20°C prior to prechilling at 3°C promotes germination (Suszka, 1967). Repetitions of the warm and cold stratification cycle result in full germination of seeds (Michalska and Suszka, 1980).

Storage Conditions

Cherry seeds are orthodox and remain viable for longer periods at low-moisture and low-temperature conditions. Seeds with 12 percent mc stored at 3 to 5°C preserved viability for five years. Seeds stored in hermetically sealed bottles at –3 to –5°C maintained 91 to 97 percent germination for seven years. Seeds lost viability rapidly when stored in sealed bottles at 20 to 25°C (Solovjeva, 1966). Subsequently, high seed viability was preserved for 15 years at −5°C (Solovjeva, 1978). Storage of dry seeds (4 to 6 percent mc) in moisture-proof containers at subzero temperature (− 20°C) preserved viability for a longer period (Ellis and Hong, 1985).

REFERENCES

Cromarty, A.S., Ellis, R.H., and Roberts, E.H. 1982. *The design of seed storage facilities for genetic conservation.* Rome: International Board for Plant Genetic Resources, p. 96.

Deol, I.S., Chopra, H.R., and Grewal, S.S. 1993. Studies on seed germination and seedling growth of Sharbati peach *(Prunus persica). Punjab Hort. J.* 33:58-62.

Ellis, R.H. and Hong, T.D. 1985. *Prunus* seed germination and storage. In *Long-term seed storage of major temperate fruits.* Rome: International Board for Plant Genetic Resources, pp. 3-21.

Giersbach, J. and Crocker, W. 1932. Germination and storage of wild plum seeds. *Boyce Thompson Inst. Contrib.* No. 4:39-51.

Michalska, S. 1982. Embryonal dormancy and induction of secondary dormancy in seeds of Mazzard cherry *(Prunus avium). Arboretum Kornickie* 27:311-332.

Michalska, S. and Suszka, B. 1980. The effect of multiple induction of dormancy in *Prunus avium* seed. In *Secondary dormancy of seeds of* Prunus *species.* Polish Acad. Sci. Inst. Dendrology, Kornik, Poland, pp. 13-24.

Solovjeva, M.A. 1966. Long term storage of fruit seeds. *Proc. XVIIth Inter. Hort. Congr.* 1:238.

Solovjeva, M.A. 1978. Method of storing fruit seeds and determining their storage quality. *Bull. Vses. Ord. Len. Inst. Rast, Vavilova* 77:64-65.

Suszka, B. 1964. The influence of method and duration of stone storage on the germination capacity of mazzard cherry *(Prunus avium* L.). *Arboretum Kornickie* 9:223-235.

Suszka, B. 1967. Studies on dormancy and germination of seeds, from various species of the genus *Prunus. Arboretum Kornickie* 12:221-282.

Toit, H.J., Jacobs, G., and Strydom, D.K. 1979. Role of the various seed parts in peach seed dormancy and initial seedling growth. *J. Am. Soc. Hort. Sci.* 104:490-492.

30

Raspberry (*Rubus* spp.)

Introduction

Raspberry is a hardy plant widely cultivated in temperate areas of the northern hemisphere. Red *(Rubus idaeus)*, black *(R. occidentalis)*, and purple raspberry (hybrid of red and black) are cultivated in different locations. Fruits are eaten fresh, canned, and used in the preparation of jam, jelly, syrup, and wine. Leaves are used for medicinal purposes. Raspberry is native to North America, Asia, and Europe. It is predominantly cultivated in England, Hungary, Russia, the United States, and Yugoslavia.

Morphology

Raspberry belongs to family Rosaceae, and its diploid chromosome number is 2n = 14. Plant is erect and about 1 to 2 m in height. Suckers are commonly used for its propagation. Seeds are used for evolving new cultivars through hybridization. Flowers are small, few, and white. Fruit is oblong, conical, and dark red. Ripe and healthy fruits are selected for seed purposes. Seeds are extracted from the fruit, washed thoroughly with water, dried, and packed for storage.

Seed Storage

Seed Germination

Fresh raspberry seeds show poor and delayed germination, attributed to the presence of a hard seed coat (Marcuzzi and Fernandiz, 1993). Seed scarification and removal of the seed coat hasten the germination process. Jennings and Tulloch (1965) reported that raspberry seeds germinated fully when scarified seeds were exposed to prechilling, alternate temperatures, or light or treated with potassium nitrate, thiourea, or kinetin. Further, gibberellins, sodium hypochlorite, calcium hypochlorite, and prolonged warm stratification promote the germination of intact seeds. Seed treatment with potassium hydroxide, sulfuric acid, or gibberellic acid stimulates germination. Seed germination was higher with potassium hydroxide, and seeds soaked in concentrated sulfuric acid for more than 20 min were damaged (Misic and Belic, 1973).

Storage Conditions

Raspberry seeds show orthodox storage behavior. Low seed moisture and low storage temperatures are beneficial for extending storage life of seeds. Seeds stored in polyethylene bags at 1°C remained viable for three months (Misic and Belic, 1973).

REFERENCES

Jennings, D.L. and Tulloch, B.M.M. 1965. Studies on factors that promote germination of raspberry seeds. *J. Exp. Bot.* 16:329-340.

Marcuzzi, M.I. and Fernandiz, De.M.E.A. 1993. Study of dormancy in seeds of raspberry (*Rubus idaeus* L.). *Phyton.* 54:139-147.

Misic, P.D. and Belic, M.V. 1973. Methods for improving the germination of red raspberry seeds. *Jugoslovensko Vocarstvo* 7:153-156.

31

Blueberry (*Vaccinium* spp.)

Introduction

Blueberry is an important small fruit in North America and Europe. It is cultivated for the edible fruits. Fruits are consumed fresh as well as processed. Products such as jam, jelly, sauce, pie, and wine are prepared. Plants are also used for ornamental purposes in home gardens. Fruits are a rich source of sugar, citric acid, vitamin C, phosphorus, calcium, iron, and potassium.

Blueberry is native to North America. It is cultivated predominantly in Bulgaria, Canada, Chile, Denmark, Finland, Japan, the Netherlands, New Zealand, Poland, Russia, and the United States.

Morphology

Blueberry belongs to family Ericaceae and has chromosome number 2n = 24. The commonly cultivated blueberries are highbush (*Vaccinium corymbosum*), lowbush (*V. angustifolium*), and rabbiteye (*V. ashei*) blueberry. The plant is a deciduous shrub, growing up to 6 m in height. Flowers are small, borne in clusters, and white or tinged with pink. Fruit is blue and sweet. Seeds, grafting, and cuttings are used to propagate blueberry bushes. Lowbush and rabbiteye blueberries are self-incompatible and require cross-pollination for higher fruit set (Stushnoff and Palser, 1969). These are mainly pollinated by honeybees. Fruit becomes blue on full ripening. Fruit with larger seeds develops faster than fruit with smaller seeds. Healthy and fully ripe fruits are collected for seed purposes. Seeds are removed, washed thoroughly with water, dried in shade, and packed in moisture-proof containers for storage.

Seed Storage

Seed Germination

Blueberry seeds germinate readily on immediate extraction from ripe berries. Seed stratification at 3 to 4°C for 30 days promotes seed germination (Butkene and Butkus, 1980). Seed germination was higher when berries were stored at $-23°C$ and seeds were stored in berries at $-2°C$ or $1°C$. The berries become moldy with *Botrytis* spp. and germination percentage is reduced considerably (Aalders and Hall, 1975). Mature and large seeds give higher germination than immature and small seeds. Seed soaking in gibberellic acid (100 to 500 ppm) stimulates the germination process (Ballington, Galleta, and Pharr, 1976). Light also hastens the germination process in fresh seeds (Butkene, 1989).

Storage Conditions

Blueberry seeds preserve well under ambient conditions. High seed viability is preserved for two years at 5 to 7°C (Butkene, 1989). Further, seeds preserved at 4.4°C gave higher germination (86.2 percent) after 12 years of storage (Darrow and Scott, 1954).

REFERENCES

Aalders, L.E. and Hall, I.V. 1975. Germination of lowbush blueberry seeds stored dry and in fruit at different temp. *HortSci.* 10:525-526.

Ballington, J.R., Galleta, G.J., and Pharr, D.M. 1976. Gibberellin's effects on rabbiteye blueberry seed germination. *HortSci.* 11:10-11.

Butkene, Z.P. 1989. Biological and biochemical characteristics of highbush blueberry. *5. Seed germination. Biologijos Mokslari* No. 2:38-44.

Butkene, Z.P. and Butkus, V.F. 1980. Biological and biochemical characteristics of blueberry. 2. Effect of the duration and method of stratification, time of seed harvest, temperature, light, substrate and medium pH on seed germination. *Trudy Akad Nauk LitSSR* No. 3/91:45-55.

Darrow, G.M. and Scott, D.H. 1954. Longevity of blueberry seed in cool storage. *Proc. Am. Soc. Hort. Sci.* 63:271.

Stushnoff, C. and Palser, B.F. 1969. Embryology of five *Vaccinium* taxa including diploid, tetraploid and hexaploid species, or cultivars. *Phytomorphology* 19:312-321.

32

Walnut (*Juglans* spp.)

Introduction

Walnut is native to Iran and the surrounding areas. It is extensively cultivated in the temperate regions of Bulgaria, China, France, Iran, Italy, Mexico, Romania, Turkey, and the United States, where summers are relatively warm. All plant parts of walnut are used. Immature fruits are used in the preparation of pickles, marmalades, and juice. Walnut is rich in proteins, fats, and minerals, such as phosphorus, potassium, and magnesium, and in vitamin B. Walnut oil is used in cosmetics. Trees are planted for landscaping and ornamental purposes. Its wood is highly valued for making various types of furniture.

Morphology

Walnut belongs to family Juglandaceae and has somatic chromosome number 2n = 32. The various types of cultivated walnut are Persian walnut *(Juglans regia)*, black walnut *(J. nigra)*, and Heartseed walnut *(J. ailanthifolia)*. Trees are large, deciduous, monoecious, aromatic, ornamental, and long-lived. They grow up to 20 to 25 m in height. Nuts are enclosed within thick indehiscent husks. Walnut is pollinated by the wind. It is commonly propagated by seedling and vegetative means such as grafting and budding, and stooling is also employed. Seedlings are largely raised for rootstock purposes. It takes eight to ten years for fruiting. Hulls develop a crack on maturity. Healthy and mature nuts are collected for seed storage.

Seed Storage

Seed Germination

Walnut seed germination is slow and takes longer to complete. Seeds are dormant and need stratification to break the dormancy (Sharma and Chauhan, 1981). Removal of a small part of the hard shell promotes radicle emergence.

Exposure of seed to longer chilling periods overcomes dormancy (Gordon and Rowe, 1982). Seed immersion in hot water (80 to 90°C) for 24 h followed by stratification for a short period results in uniform germination (Molchenko et al., 1977). Seed treatment with ethephon (100 ppm) improves the germination percentage (Casini and Salvadori, 1975).

Storage Conditions

Walnut seeds show orthodox storage behavior (Gordon and Rowe, 1982). Seed longevity increases when seeds are stored under dry, cool storage conditions. Seed viability is higher (37 percent) in seeds stored loose in pots after four years of storage. In damp conditions (50 percent mc and 2.8°C) germination is less than 20 percent, but at lower seed moisture (30 percent) germination is as high as 76 percent (Williams, 1971).

REFERENCES

Casini, E. and Salvadori, S. 1975. Observations and research on the use of various growth regulators in the germination of seeds of fruit trees. *La Nuova AOPI*, p. 22.
Gordon, A.G. and Rowe, D.C.F. 1982. Seed manual for ornamental trees and shrubs. *Forestry Commission Bull.* 59:132.
Molchenko, L.L., Shevchuk, N.S., Yatsyshin, V.M., and Krushelnitskii, A.F. 1977. Rapid preparation of *Juglans nigra* seeds for sowing. *Ref. Zhurnal* 6.55.1025.
Sharma, S.D. and Chauhan, J.S. 1981. Effect of shell thickness on seed germination and growth of seedlings obtained from the nuts of seedling walnuts. *South Indian Hort.* 29:87-89.
Williams, R.D. 1971. Storing black walnut seeds. *62nd Ann. Rep. Northern Nut Growers Assoc. Inc. Carbondale, Illinois*, pp. 87-89.

SECTION III:
SEED STORAGE
IN VEGETABLE CROPS

33

Potato: *Solanum tuberosum* L. and Sweet Potato: *Ipomoea batatas* (L.) Lam.

POTATO

Introduction

Potato is a vegetable crop with global importance. It has been acknowledged as a staple food crop in many countries and ranks fourth after rice, wheat, and maize in production. It is cultivated in temperate, subtropical, and tropical regions for its underground edible tubers. Potato is a good source of carbohydrates and proteins and also contains high quantities of potassium and vitamins such as niacin, thiamine, and riboflavin. It has industrial importance in the manufacturing of starch, and also in the making of potato chips.

Origin and Distribution

Potato is native to South America. Its genetic diversity in terms of landraces and wild species occurs in Bolivia, Colombia, Ecuador, Peru, and Venezuela. Potato cultivation is distributed between 40°N and 20°S latitudes. It grows well in temperate regions, especially in Europe and North America. In the tropics, it does better at higher elevations. Besides North America, it is cultivated in Belarus, China, France, Germany, India, Japan, Kenya, Malaysia, the Netherlands, the Philippines, Poland, Russia, Sudan, Ukraine and United Kingdom.

Morphology

Potato is an important member of the Solanaceae family. Plants are a dicotyledonous, perennial herb that is cultivated as an annual; the food is stored in tubers at the end of the stolons. Tubers vary in size, shape, color, and depth of eyes, which are characteristics of the different cultivars. They are bulky, perishable, and difficult to transport and store for long periods. Potato stems are weak; leaves are alternate and pinnately compound with ovate leaflets. The cultivated potato is tetraploid (2n = 48) and diploid (2n = 24), while hexaploid (2n = 72) potatoes are also found. Flowers are small and white, yellow, blue, or purple and are borne in clusters. Fruit is a berry, globose, green or purplish green, and not edible (see Figure 33.1). It contains more than 300 seeds. Seeds are small (1.5 mm), yellow or brown, flat or kidney shaped, and poisonous. One thousand seeds weigh roughly 0.6 g.

Seed Storage

Potato is propagated commercially by tubers, normally called seed tubers, to maintain the identity of the varietal characteristics, whereas true seeds are used in crop improvement and genetic conservation. Of late, cultivation through true potato seeds (TPS) is becoming popular in many potato-growing areas, due to nonavailability of sufficient quantity of quality seed

FIGURE 33.1. True Potato Seeds and Seed Tuber

Fruits

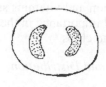

Cross Section of Fruit

Seeds

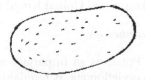

Tuber

tubers, high transport costs, and problems in tuber storage. Potato seeds show orthodox storage behavior. These tiny seeds can be dried to a low level of moisture and stored at low temperatures for fairly long periods without affecting seed quality.

Seed Collections

Potato is a cool-season crop, requiring low temperatures for optimum growth and tuberization. It is cultivated as a summer crop in temperate regions but is grown during the winter in the tropics and subtropics. It needs well-drained sandy loam soils. Soils rich in nutrients such as nitrogen, phosphorus, and potassium are favorable for high crop production. Potato flowers profusely and sets more fruits (berries) during long days. Both self- and cross-pollination occur, but many cultivars do not produce a sufficient quantity of pollen. However, a few set moderate to abundant amounts of pollen. Normally, potato is hand pollinated twice for better seed setting (Thakur and Upadhya, 1993). Fruit matures in six to seven weeks after anthesis. Ripe healthy fruits are harvested for seed purposes, and seeds are separated, cleaned, dried, and stored.

Seed germination. Seed germination is epigeal and takes three to four weeks to complete. Seed germination is better at a constant temperature of 20°C and under diffuse light. Seed coat removal and exposing seeds to 16 to 118°C for 12 days increases germination percentage (Subrahmanyam, 1971). Fresh seeds exhibit dormancy, which can be removed by after-ripening of seeds for 6 to 24 months at room temperature (Simmonds, 1963).

Seed dormancy. Fresh seeds from fully developed berries show prolonged dormancy, and seeds of early harvested berries are less dormant and show poor viability. Dormancy is beneficial, especially in delaying the germination process, and enables longer storage of seeds under ambient conditions. Seeds of tetraploid plants are more dormant than seeds of diploid plants (Simmonds, 1963). Dormancy can be overcome by storing seeds for seven months. The endogenous gibberellic acid level increases during the first 120 days of storage, and simultaneously the quantum of abscisic acid (ABA) decreases. The proportion of GA and ABA regulates seed dormancy (Bhargava, 1997). However, the residual dormancy persists for two to eight years in potato seeds (Antonio and Mcttale, 1988). The external application of gibberellic acid (1,000 ppm) removes seed dormancy (Sathyamoorthy and Nakamura, 1995).

Storage Conditions

Potato seeds are viable for relatively long periods, and they do not require specialized and expensive storage structures, as in the case of tubers

(Gaur and Pandey, 1993). Seeds require dry, cool conditions for maintaining high seed quality in terms of viability and vigor. Seeds with 5 to 7 percent mc stored at 6 to 10°C over calcium chloride remain viable for eight years (Thakur and Upadhya, 1990). Further, about 70 percent of seeds are viable even after ten years of storage at fluctuating temperatures (5 to 45°C) (Subrahmanyam, 1971). High seed viability is maintained for 10 to 13 years when seeds are stored at 0°C (Clark, 1940). At lower temperatures (0 to 5°C), seeds remain dormant for 5 to 13 years (Clark, 1940; Simmonds, 1968). Seeds that are low in moisture when stored in sealed containers at 1.1°C maintain high viability and vigor for nine years (Stevenson and Edmundson, 1950).

SWEET POTATO

Introduction

Sweet potato is also called Spanish potato. It is cultivated throughout the tropics and subtropics for its edible tuberous roots. Tubers are rich in carbohydrates and beta-carotene, and they contain a high percentage of sugars, predominantly sucrose, about 3 to 6 percent. Tubers are boiled, baked, fried, and canned. Potato flour is used in the preparation of starch, glucose, pectin, and industrial alcohol. The tender foliage, which is rich in vitamins A and C, is used as a leafy green vegetable.

Origin and Distribution

Sweet potato originated from *Ipomoea trifida* in tropical South America (Nishiyama, 1971). It is largely cultivated in Argentina, Brazil, Columbia, Ecuador, India, Indonesia, Japan, Malaysia, South Korea, and the Philippines.

Morphology

Sweet potato belongs to family Convolvulaceae and has a chromosome number of 2n = 90. It is a hexaploid, dicotyledonous perennial plant with long trailing vines that is cultivated annually. Roots are tuberous, fusiform, or globular, and there are four to ten per plant. Tubers vary in size and color, ranging from white, tan, and yellow-orange to salmon orange or red. Leaves are alternate and vary in shape. Sweet potato flowers readily in the tropics. Flowers are single and axillary; the calyx is five lobed; the corolla is funnel shaped and purple in color; stamens number five; the ovary is bilocular. Fruit is a dehiscent capsule that contains one to four seeds. Seeds are black, angular, and 3 mm long with hard testae. A large number of seeds set when tops are grafted on roots of cultivated morning glory plants.

Seed Storage

Stem cuttings are normally employed for propagation. Seeds are largely used in breeding programs and for conservation of genetic diversity in the gene bank. Seeds are stored for fairly long periods under ambient conditions. They show orthodox storage behavior, can tolerate desiccation, and store well at chilling temperatures.

Seed Collections

Sweet potato is a warm-season crop. It grows well in hot climates and needs plenty of sunshine. Low temperatures (less than 15°C) affect plant growth. Short days are congenial for flowering and root development. Sweet potato requires deep, friable, sandy loam soil. The soil should be well drained. It is cultivated through root and vine cuttings (30 to 45 cm long). Later, it is transplanted on ridges in the main field. Crops should be provided with adequate nutrition, irrigation, and plant protection measures for healthy and optimum growth. Sweet potato is self-incompatible, and cross-pollination is brought about by insects. Matured dry fruits are collected for seed purposes and are extracted for storage.

Seed Germination

Sweet potato seeds possess a hard seed coat that delays the imbibition process. Therefore, seed scarification with concentrated sulfuric acid is helpful in encouraging early and increased seedling emergence.

Storage Conditions

In sweet potato, seed deterioration is greater under high seed moisture and high storage temperatures. However, low-moisture seeds stored in moisture-proof containers under cool conditions retain viability for longer periods. Jones and Dukes (1982) stored seeds at 18°C and 45 to 50 percent RH for 21 years, and 90 percent of sweet potato seeds germinated.

REFERENCES

Antonio, V.L.D. and Mcttale, N.A. 1988. Effect of storage on germination of true potato seeds. *Am. Potato J.* 65:573-581.

Bhargava, R. 1997. Changes in abscisic and gibberellic acids contents during the release of potato seed dormancy. *Biol. Plant.* 39:41-45.
Clark, C.F. 1940. Longevity of potato seeds. *Am. Potato J.* 17:147-152.
Gaur, P.C. and Pandey, S.K. 1993. True potato seeds. In *Advances in horticulture,* Volume 7, Chadha, K.L. and Grewal, J.S. (Eds.). New Delhi: Malhotra Publishing House, pp. 85-111.
Jones, A. and Dukes, P.O. 1982. Longevity of stored seeds of sweet potato. *Hort. Sci.* 17:756-757.
Nishiyama, I. 1971. Evolution and domestication of the sweet potato. *Bot. Mag. Tokyo* 84:377-387.
Sathyamoorthy, P. and Nakamura, S. 1995. Effect of gibberellic acid and inorganic salts on breaking dormancy and enhancing germination of true potato seeds. *Seed Res.* 23:5-7.
Simmonds, N.W. 1963. Experiments on germination of potato seeds. *Euro. Potato J.* 6:69-76.
Simmonds, N.W. 1968. Prolonged storage of potato seeds. *Euro. Potato J.* 11:150-156.
Stevenson, F.J. and Edmundson, W.C. 1950. Storage of potato seeds. *Am. Potato J.* 27:408-411.
Subrahmanyam, K.N. 1971. Germination of the potato seeds under long storage conditions. *Curr. Sci.* 20:379-380.
Thakur, K.C. and Upadhya, M.D. 1990. True potato seeds production technology. In *Commercial adoption of true potato seed technology: Prospects and problems,* Gaur, P.C. (Ed.). Shimla: Central Potato Research Institute, pp. 19-28.
Thakur, K.C. and Upadhya, M.D. 1993. Technology for hybrid true potato seed production in India. *Seed Res.*1:82-93.

34

Eggplant: *Solanum melongena* L., Tomato: *Lycopersicon esculentum* Mill., and Peppers: *Capsicum annuum* L.

EGGPLANT

Introduction

Eggplant is one of the most popular vegetable crops in all the tropical and subtropical countries. Eggplant is also called brinjal or aubergine. The name eggplant is derived from the egg-shaped fruit. It is a popular vegetable cultivated for its edible fruits. The immature fruit is cooked as a vegetable, fried, stuffed, or boiled.

Origin and Distribution

Eggplant is native to India, where wide genetic diversity exists. Its wild type and landraces, which are distributed in large areas, are spiny and bitter. It is largely cultivated in China, Egypt, Ghana, India, Indonesia, Italy, Iraq, Japan, Kenya, Malaysia, Nigeria, the Philippines, Puerto Rico, Spain, Syria, Thailand, and Turkey.

Morphology

Eggplant belongs to family Solanaceae, and its chromosome number is 2n = 24. It is a weak, erect, branching, spiny, pubescent, dicotyledonous, perennial herb. It is normally cultivated as an annual. Plants are moderately tall, growing up to 1 m in height. Foliage is dark green with large thick ovate or ovate-oblong leaves. Flowers are violet colored, single or in clusters of two to five. The calyx is long, woolly, often spiny, and it is persistent on the fruit even after maturity. Petals number five and are gamopetalous; stamens number five, are yellow, and stand erect. The ovary is superior and bilocular. Fruits vary in size, shape, and color and have many seeds. They are oblong or obovoid; smooth and shiny; and yellow, green, white, purple, or black. Seeds are small and numerous, kidney shaped, and light brown in color. One thousand seeds weigh about 4 g.

Seed Storage

Seeds are used for commercial cultivation of the crop. They are also used in genetic conservation, owing to ease of handling, are capable of maintaining genetic stability, and are inexpensive. Seeds show orthodox storage behavior. They require dry, cool conditions for better storage.

Seed Collections

High viability and vigor are ideal qualities for seed storage. These help to withstand unfavorable storage conditions. Vigorous seeds remain viable for longer periods under both favorable and unfavorable storage conditions. Seeds should be bold, clean, and free from debris as well as from insects and other microorganisms for better storage. Quality seeds are thus produced when optimum cultural practices are followed during seed production, and these seeds can then be collected for storage.

Eggplant is a warm-season crop and therefore needs high temperature for optimum growth. Variations in temperature affect plant growth. The optimum soil temperature for seed germination is 30°C. Comparatively, seeds of long-fruited cultivars better withstand extremes of heat. The crop grows best in rich, deep, well-drained, sandy loam soils. Seeds are sown in raised nursery beds. Healthy seedlings are transplanted in the field after four to six weeks. In high rainfall areas, crops are grown on raised beds. Optimum spacing, fertilization, irrigation, and other cultural practices should be followed. Eggplant is a self-pollinated crop. However, cross-pollination occurs to a certain extent. Thus, an isolation distance of 50 m is provided to safeguard against contamination. Plants are checked for being true to type, especially during the flowering and fruiting stages, and any off types are removed. Fruits are harvested about 90 days after planting. Fruit color changes on maturity and its glossy appearance is lost. Simultaneously, seeds mature and become ready for harvest. Seeds are extracted from dry ripe fruits either manually or by wet extraction, whereby fruits are crushed and the pulp and seeds are soaked in water overnight. Later, the seeds are separated, dried, cleaned, and suitably packed for storage.

Seed Germination

Fresh seeds of eggplant do not germinate satisfactorily, requiring more time for germination, and thus show a certain amount of dormancy (Suzuki and Takahashi, 1968). Dormancy can be overcome by partially soaking seeds in potassium nitrate (4 percent) and subjecting them to an alternate temperature (20/30°C) regime. Seed treatment with gibberellic acid (100 ppm) eliminates dormancy (Krishnasamy and Rangarajpalaniappan, 1990).

Seed germination is epigeal and takes two weeks to complete. Germination is better at a constant temperature of 25°C or at alternate temperatures of 20/30°C. Kretschmer (1981) reported that seeds stored at room temperature or under low temperature for three years germinated well at 15 to 30°C, while Suzuki and Takahashi (1968) suggested that eggplant seeds require alternate temperatures for full germination. Temperatures of 30/23°C for 8/16 h, respectively, are sufficient to promote germination in fresh and one-year-old dormant seeds (Winden and Bekendam, 1975). Seed germination is higher in heavier seeds than in lighter ones, and this is attributed to better seed filling (Selvaraj, 1988). Seed weight is positively correlated to protein content and, in turn, to seedling vigor. Further, higher seedling vigor from heavier or larger seeds is attributed to better availability and metabolism of reserve food (Pollock and Roos, 1972). Eggplant seeds germinate better under dark conditions (Quagliotti and Rota, 1986). However, seeds are sensitive to salt, and germination is affected in saline soils (Cucci et al., 1994).

Factors Affecting Seed Longevity

Eggplant seeds can be stored fairly well under ambient conditions. High seed moisture at higher temperature affects seed longevity.

Genetic factors. Seed longevity varies within the genus among the different species. Certain cultivars withstand unfavorable storage conditions and remain viable for limited periods. Such cultivars also do not require special conditions for short-term storage (Doijode, 1992).

Fruit maturity. Eggplant fruits take 30 to 40 days to mature following anthesis. Seeds extracted from fully ripe fruits give high seed quality, as seed maturity coincides with that of fruit. Mature seeds retain viability better than immature seeds (Eguchi and Yamada, 1958).

Fruit storage. Eggplant fruits are stored for a week under ambient conditions. Further storage causes shriveling of fruits, affecting seed quality. Seeds from underripe fruits are particularly susceptible to freezing temperatures and become dormant; warming improves germination. Storing fruits at low temperatures or below 0°C reduces germination (Alekseev, 1976). Storage of mature fruits for seven days increases the germination percentage (Petrov and Dojkov, 1970).

Seed position in fruit. Seeds from the basal portion of fruits give early and high germination. These seeds preserve their high quality on storage (Doijode, unpublished data). Similarly, fruits borne on the lower portion of the plant give quality seeds and higher germination (Petrov and Dojkov, 1970). Size of fruit and seed weight are reduced in later-formed fruits (Naik et al., 1995).

Storage pests. Quite a few insects and fungi are associated with storage of eggplant seeds. Many of the pests can be eliminated under drier storage conditions. Drugstore beetle (*Stegobium paniceum* L.) causes considerable seed damage and affects seed viability. It is controlled by fumigation with methyl bromide. The viral activity inside the seed is entirely lost after seven months of storage at room temperature, without an appreciable reduction in germination (Mayee, 1977). Fungi are controlled by a combination of activated clay and carbendazim. Application of thiram or Delsan-30 protects the seeds over a period of 18 months, without affecting the germination (Reddy and Reddy, 1994). Seed dressing with organomercurial compounds severely reduces seed viability (Nakamura, Sato, and Mine, 1972).

Storage conditions. Seed longevity is primarily dependent on temperature, relative humidity, and, to a lesser extent, oxygen. Extremes of temperature and relative humidity result in a rapid decline in seed viability and seedling vigor, and an increase in leaching of electrolytes (see Figure 34.1) (Doijode, 1988b). Seed deterioration and invasions by fungi are greater at higher levels of seed moisture. The moisture is regulated by atmospheric humidity under open storage. Barton (1943) reported that seed viability was affected by high humidity during storage (76 percent RH for 64 weeks). Seeds stored at relative humidity of 10 to 40 percent maintained 50 to 60 percent viability after five years of storage under ambient conditions (Doijode, unpublished data). Seed moisture content of 5 to 7 percent is ideal for long storage.

Storage Methods

Well-dried eggplant seeds maintain high viability under cooler conditions. Seeds are protected from extraneous factors, including humidity, by providing suitable moisture-proof packaging. Polyethylene bags are ideal for short storage, whereas laminated aluminum foil pouches are effective for longer storage.

Seed storage with silica gel. Seed storage with silica gel is beneficial in lowering the moisture content, thereby prolonging storage life. Silica gel is an inactive, inexpensive, and useful self-indicator material for seed storage. Seeds stored with silica gel in aluminum foil under ambient conditions recorded 50 percent germination after 30 months of storage, as compared to seeds stored without silica gel, which had 50 percent germination at 12 months of storage (Thulasidas, Selvaraj, and Thangaraj, 1977).

Cold storage of seeds. Seed viability and vigor preserve well under low temperatures. Mohamed, Lester, and Mumford (1988) reported that seed viability declines rapidly under warm, humid conditions. While seeds with

FIGURE 34.1. Influence of High Temperatures (42°C) and Humidity (90 Percent RH) on Seed Viability, Vigor, and Leakage in Eggplant

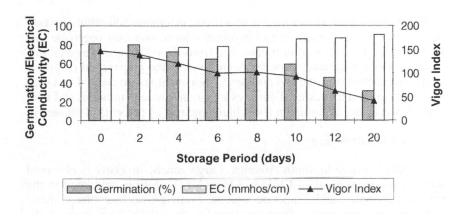

7 percent mc maintain viability of 92 to 98 percent at –18 to 20°C for 112 days, seeds stored in laminated aluminum foil pouches at 5°C and –20°C maintain high viability and vigor for ten years (Doijode, unpublished data). Viability decreases to 50 percent during the second year of ambient storage (Doijode, 1998).

Seed storage in modified atmosphere. High atmospheric oxygen is injurious to seed viability, especially at higher levels of seed moisture. Eggplant seeds stored in laminated aluminum foil pouches along with carbon dioxide exhibited high seed germination after four years of storage under ambient conditions (Doijode, unpublished data).

Invigoration of Stored Seeds

Seed deterioration is a gradual and irreversible process resulting in loss of seed quality. Imposing certain treatments slows down this process. In one-year-old seeds, germination is improved by naphthaleneacetic acid (NAA) (100 ppm) application (Suryanarayana and Rao, 1984). Similarly, in five-year-old eggplant seeds, germination and vigor are improved by seed priming with gibberellic acid (50 ppm) and potassium nitrate (0.01 M). (Demir, Ellialtioglu, and Tipirdamaz, 1994). Midstorage application of disodium phosphate to eggplant seeds is shown to improve seed germination, crop growth, and yield (Geetharani, Ponnuswamy, and Raju, 1996).

TOMATO

Introduction

Recent years have witnessed a dramatic rise in tomato cultivation in many areas of the world. Tomato is an important solanaceous fruit vegetable that ranks fifth in production after potato. The fruits are edible and are eaten raw or cooked as a vegetable. Ripe fruits are used in the preparation of puree, soup, juice, ketchup, paste, and powder. Green tomatoes are cooked as a vegetable and made into pickles. Tomatoes are a rich source of minerals and vitamins A and C. Seeds contain a fairly high amount (24 percent) of edible oil.

Origin and Distribution

Tomato is native to South America. Large genetic diversity is observed within the coastal areas of Chile and Ecuador. Tomato is cultivated in the tropics, subtropics, and temperate regions. Fruits grown in cooler climatic conditions possess greater flavor and quality than those from tropical climates. Tomato is largely cultivated in Algeria, Brazil, China, Egypt, Greece, India, Indonesia, Iran, Italy, Japan, Malaysia, Mexico, Morocco, the Philippines, Spain, Turkey, and the United States.

Morphology

Tomato belongs to the Solanaceae family and has chromosome number $2n = 24$. Plant is weak and erect or climbing; it shows indeterminate, semideterminate, or determinate type of growth. It grows up to 2 m in height, and branches heavily with compound leaves. Leaves are dark green and hairy, with a characteristic odor. Flowers are yellow, perfect, and borne in clusters of four or more. Flower is composed of a short calyx tube; corolla rotate with six petals; six stamens that form on the corolla tube; and a pistil with several locules (five to nine). Fruit is a soft fleshy berry with various shapes, sizes, and colors. It may be round, pear, oblate, somewhat square, or flattened, with a smooth or ribbed surface. The skin color changes to red or yellow on ripening, but it remains green in fruits of *Lycopersicon peruvianum, L. chilense,* and *L. hirsutum.* Seeds are many, small, light brown, hairy, and kidney shaped, with a curved embryo embedded in the endosperm. One thousand seed weigh about 2.5 to 3.3 g.

Seed Storage

Tomato seeds are tiny structures capable of withstanding adverse storage conditions, to a certain extent. They show orthodox storage behavior, wherein desiccation of seeds and storage at low temperatures increase seed longevity. Seeds are primarily used in propagating tomato and in breeding and genetic conservation. Farmers prefer to preserve viable seeds for short periods, such as until the next growing season, whereas breeders demand the fairly long storage for germplasm conservation. Therefore, suitable storage methods have been developed to suit these requirements.

Genetic Factors

Tomato seeds maintain their viability for long periods, and this is attributed to genetic factors (James, Bass, and Clark, 1964). Tomato seed longevity is primarily controlled by genetic means, and the magnitude of viability and germination potential is higher in hybrids than in inbred plants (Palaniappan, Muthukrishnan, and Irulappan, 1981). Seed longevity differs within the genus among species and cultivars. Seed germination percentage varies among different cultivars stored for four years under ambient conditions (Doijode, 1987b).

Seed Collections

Tomato is a warm-season crop and cannot survive at very hot or very cold temperatures. It prefers deep, fertile, well-drained soil. The optimum soil temperature for germination is 30°C, with a minimum temperature of 10°C. Seeds are sown in nurseries or on raised beds or ridges, and after four to six weeks, the seedlings are transplanted to the field. Seedlings are supported with stakes, especially the indeterminate types. Proper fertilization, timely irrigation, and plant protection measures should be followed. Tomato is a self-pollinated crop. An isolation distance of 50 m is maintained to avoid accidental cross-pollination by insects. Plants are examined for cultivar characteristics during the flowering and fruiting stages, and the off types are removed. It takes about 60 to 90 days from planting to harvesting, depending on cultivars. Extreme temperatures, especially higher than 38°C, affect the fruit set. Ripe fruits are harvested (see Figure 34.2). Seeds are removed by cutting the fruits; they are then washed in water and dried. Pulp is also separated by fermentation or by using dilute hydrochloric acid. Normally, medium to large fruits are picked for seed purposes. Seeds of these fruits give higher germination and vigorous seedlings (Yaovalak, 1987).

Fruit maturity. Tomato seeds reach acceptable quality on maturity of fruits. The seeds mature between 35 and 41 days after anthesis, at which time seed moisture will be about 53 to 72 percent. Seeds taken 21 days after flowering are capable of germinating but do not store long (Demir and Ellis, 1992a). Seed germination, vigor, and storage protein increase, and seed

FIGURE 34.2. Tomatoes Ready for Harvest

moisture decreases, with advances from mature green to the ripening stage (Singh and Lal, 1990). Fruit color changes on maturity, when fruit becomes dark red and seed dry weight and vigor reach the maximum. Yaovalak (1987) noted that seed germination and vigor are unaffected in fruits showing pale pink to dark red coloring. Seeds of mature green fruit give low germination, and rate and intensity of germination increase from the color breaker stage. Seeds attain maximum viability in the ripe stage of fruit development and show greater seedling vigor. However, fruits can be plucked at the color breaker stage for an early seed crop without hampering seed viability (Doijode, 1983). Likewise, Singh, Singh, and Dhillon (1985) obtained 95 percent germination in seeds of fruits turning red and in red ripe fruits, as compared to only 64 percent in seeds of mature green fruits. Further, seed quality is affected in seeds of overripe and diseased fruits (Alekseev and Prokhorov, 1982).

Fruit storage. Storing fruit for a short period provides ample time for preparation and processing of seed material for storage. Fruits of different stages retain good seed quality for various periods. Seed viability decreases with an increase in fruit storage time. Red ripe tomatoes stored in a refrigerator (11.5°C and 57 percent RH) retain seedling vigor for six weeks and show the highest vigor during the third week of storage, as compared to the low vigor shown when stored for a short period at room temperatures (27.7°C and 75 percent RH) (Baldo et al., 1988).

Seed drying. Tomato seeds have 60 to 70 percent mc at harvest, and this must be reduced to 9 percent during storage for better seed quality (Harrington, 1960). Seeds dried in sun and shade (alternately for 2 h) and mechanical drying with air at 40°C are optimal for safe drying of seeds for better storage.

Seed Germination

Seed germination is epigeal and takes about one week to complete at higher temperatures. The optimum temperature is from 25 to 30°C (Mancinelli, Borthwick, and Hendricks, 1966). Tomato seeds show considerable dormancy (Shuck, 1936). Secondary dormancy is induced by chilling (Went, 1961), warm stratification (Mobayen, 1980), and exposure to far red light (Mancinelli, Borthwick, and Hendricks, 1966). Seed priming is reported to improve the germination percentage (Cavallaro, Mauromicale, and Vincenzo, 1994).

Seed size. Medium-sized seeds are reported to give higher germination, and seed size of about 0.8 mm is found to give higher field emergence (Pandita and Randhawa, 1995) and higher yield (Kartapradja, 1988).

Plant extracts. Plant extracts contain certain growth substances that either promote or inhibit the germination process. Tomato juice inhibits germination, an effect attributed to the presence of abscisic acid (Santos and Yamaguchi, 1979). Similarly, tomato seed germination is inhibited by mandurin juice (Bhutt and Mazumdar, 1975) and *Emblica officinalis* fruits (Ponappa and Mazumdar, 1976). Plant extracts of *Ceratophyllum demersum, Eichhornia crassipes, Marsilea minuta, Salvinia natans,* and *Spirodela polyrrhiza* stimulate germination, whereas extracts of *Cannabis sativa, Crinum asiaticum, Holarrhena antidysenterica, Nelumbo nucifera,* and *Sansevieria roxburghiana* inhibited germination (Tripathi and Srivastava, 1970).

Fungal extracts. Tomato seed germination is optimum in normal soils, rather than in sterilized and saline soils. Seedling growth improves with the addition of rhizosphere microflora suspension of respective plant species to the soil prior to sowing (Balasubramanian and Rangaswami, 1967). Seed coating with spores of *Aspergillus niveus, A. rugulosus, A. tamarii, Penicillium* spp., and *Trichoderma lignorum* controls pre- and post-emergence losses.

Growth substances. Seed treatment with growth substances such as GA, IAA, and NAA and short exposure to light and magnetic stimulus (Nelson, Nutile, and Stetson, 1970) enhance the seed germination process. Seed soaking in GA_{4+7} in addition to osmotic priming (−1.0 MPa polyethylene glycol [PEG] 6000) increases germination of tomato seeds (Liu et al., 1996).

Storage of germinated seeds. Pregerminated and germinated seeds remain viable for shorter periods under ambient conditions. They retain their viability for five days at 5°C. Further storage reduces viability and vigor due to the degradation of enzymes and functional structures and an increase in free fatty acids (Sunil, 1991). Likewise, pregerminated seeds could effectively be preserved in moist cheesecloth, with emergence percentage and emergence rate index unaffected (Pill and Fieldhouse, 1982). These seeds can also be preserved well in vacuum storage or in nitrogen at 7°C for 63 days and register excellent plant growth (Ghate and Chinnan, 1987). Beers and Pill (1986) preserved germinated tomato seeds for five days at 0° or 5°C by suspending the seeds in two fluid drilling gels, namely Natrosol-250 and Laponite-445.

Storage Conditions

Tomato seeds preserve well under ambient conditions (Bosewell et al., 1940). Higher temperatures and relative humidity are not favorable for storage. Seeds stored at 50°C and 77 percent RH showed slow and less germination (Harrington and Setyati-Harjadi, 1966). The optimum temperature and RH for storage are close to 0°C and not exceeding 70 percent, respectively (Kurdina, 1966). Low seed moisture, 5 to 7 percent, is ideal for longer storage, even if higher temperatures are maintained. Tomato seeds survive in outer space conditions for several years without adverse effects on germination, emergence, and fruit yield (Khan and Stoffella, 1996). Seed vigor depends on initial status of seed quality and the storage conditions. During storage, seedling vigor declines earlier and more rapidly than viability (Anonymous, 1954). Further, disintegration of cell membranes causes excessive leaching of electrolytes, soluble sugars, and free amino acids (Doijode, 1988a). The process of deterioration slows under favorable storage conditions.

Storage fungi. Good healthy seeds maintain their viability for longer periods. Seed health is affected under unfavorable storage conditions. Storage fungi are active at higher levels of seed moisture, causing discoloration of seeds. This is checked either by proper seed drying or by using suitable fungicidal treatment. Seed treatment with captan (2.5 g·kg^{-1}) is effective in controlling storage fungi, and it helps retain high viability for 18 months under ambient conditions (Jayaraj et al., 1988).

Storage Methods

Seed storage with desiccants. Seed moisture plays a major role in deterioration. Even seed exposure to high levels of relative humidity reduces viability and vigor and affects membrane integrity (Berjak and Villiers, 1973).

Desiccant-like silica gel is useful in maintaining an optimum level of seed moisture. Tomato seeds stored over silica gel preserved high viability and vigor for three years under ambient conditions (Varier and Agrawal, 1989).

Seed storage at low temperatures. Tomato seeds retained high seed viability and vigor for three years under ambient conditions. Thereafter, seed viability decreased; however, the seeds were economically viable until the eighth year of storage (Popovska, 1964). Barton and Garman (1946) reported that tomato seeds maintain viability for 13 years at room temperatures and for 18 years at –5°C, and stored seeds behaved similarly to fresh seeds. Seeds maintain high viability and vigor at low (5°C) and subzero (–20°C) temperatures for ten years (Doijode, unpublished data), as compared to four years at ambient temperatures (16 to 35°C) (Doijode, 1997a). Well-dried seeds remain viable for longer periods at lower temperatures. However, seeds should be suitably packed in moisture-proof containers to safeguard against high relative humidity and extraneous pests.

Invigoration of Stored Seeds

Seed storage under unfavorable conditions reduces the rate of germination, uniformity, and total germination and increases the number of abnormal seedlings. Such low-vigor seeds can be reactivated by physical and chemical stimuli. Prestorage humidification and seed hydropriming impart resistance to deterioration. Seed priming with PEG 8000 at 20°C for seven days increases the rate of germination and maintains high viability at lower temperatures (10 and 20°C) (Alvarado and Bradford, 1987). Owen and Pill (1994) also reported that primed tomato seeds stored at 4°C showed maximum seed viability. Prolonged high temperature (35°C) storage of primed seeds reduces the germination rate and total germination (Odell and Cantliffe, 1987). Repriming of primed seeds has some benefit but does not entirely reverse the detrimental effects of high-temperature storage (Alvarado and Bradford, 1988a). Further repriming is ineffective when seeds show greater loss of viability. However, a slight gain in early germination is reported (Alvarado and Bradford, 1988b).

Midstorage soaking of seeds in water or chemicals hastens the germination process by leaching certain toxic substances and activates the enzyme network involved in germination. Soaking tomato seeds in water followed by drying improves the germination and seedling vigor (Doijode and Raturi, 1990). Hydration is also achieved by exposing seeds to 100 percent RH for 24 h followed by redrying, promoting germination (Mitra and Basu, 1979). Petrikova (1989) soaked tomato seeds in water for 24 h followed by drying at 12°C for ten days; this resulted in higher percentage of germination, higher fruit yield, and a greater number of heavier fruits.

The beneficial effects of prestorage humidification and hydropriming related to metabolic activities induced by partial hydration; the adverse effects of osmopriming were caused by a decrease in DNA repair activity due to progression in the cell cycle (Pijlen et al., 1996). Priming enhances seed germination but simultaneously lowers resistance to deterioration. Primed seeds are vigorous but have a reduced storage life (Argerich, Bradford, and Tarquis, 1989). Vos, Kraak, and Bino (1994) reported that aging of tomato seeds involves glutathione oxidation. Priming reduces the oxidized form of glutathione. Hydration in conjunction with dilute solutions of sodium chloride (10^{-3} M) or sodium orthophosphate (10^{-4} M) improves field performance and productivity (Mitra and Basu, 1979). Seed quality of eight-month-old tomato seeds was improved by disodium phosphate (Geetharani, Ponnuswamy, and Raju, 1996), and that of one-year-old seeds by naphthaleneacetic acid (50 ppm) (Suryanarayana and Rao, 1984).

PEPPERS

Introduction

Peppers are known by several names in different regions, such as bell pepper capsicum, cayenne, chilli, paprika, red pepper, and sweet pepper, according to their shape, color, and pungency. Immature and mature fruits are eaten raw as salad vegetables, and various vegetable dishes are made from them. Sweet peppers are nonpungent, have a mild flavor, and are thus widely used in salad. Ripe chilli fruits are pungent, and their dry powder is commonly used as spice in cooking various foods. Hungarian type paprikas are more pungent, whereas Spanish and European types are less pungent. The pungency in the fruit is due to capsaicin ($C_{18} H_{27} NO_3$) content, which is mainly confined to the septa and placental tissue. Peppers are a rich source of vitamins A and C.

Origin and Distribution

Peppers originated in Central and South America, more particularly in Mexico and surrounding countries. It is widely grown in the tropics, subtropics, and temperate regions. Some of the leading pepper-growing countries are Algeria, Argentina, Bulgaria, China, Hungary, India, Indonesia, Japan, Kenya, Mexico, Malaysia, Nigeria, the Netherlands, the Philippines, Spain, Sudan, Thailand, Turkey, the United States, and Yugoslavia.

Morphology

Peppers belong to family Solanaceae, and their chromosome number is 2n = 24. They are short-lived, dicotyledonous, perennial plants in the tropics, but they are cultivated as an annual. Plants consist of a strong root and shoot system and grow up to 1.5 m in height. Plant stems are hard and branched, with simple, ovate, light green to dark green leaves. Flowers are small, axillary, and single or two to three in cluster at the node; the calyx is five lobed; the corolla is white or greenish, campanulate, and five petaled; the five stamens have bluish anthers; the ovary is two celled. Fruit is a podlike berry, erect or drooping, indehiscent, and many seeded. It varies in shape (heart, cylindrical, round, cherry, long and thin, or square), size, and color. Initially, fruits are green or purple in color and later become red, orange, yellow, brown, cream, or purplish on ripening. Seeds are somewhat circular, yellow, light in weight, and 3 to 5 mm long. One thousand seeds weigh about 3.5 to 5 g, depending on variety and growing conditions.

Seed Storage

Pepper seeds show orthodox storage behavior. Seeds are predominantly used in crop production and for genetic conservation. Seeds are genetically stable in long-term conservation. Seeds remain viable for two to three years in ambient conditions. High atmospheric humidity and high temperatures shorten the storage life of seeds. The germinating energy in stored seeds decreases more rapidly than germination capacity, and it is lost completely by three years when stored at room temperatures (Spaldon and Pevna, 1966).

The seed storage life is primarily dependent on genetic factors. Seed longevity differs among the different cultivars during storage and is influenced by storage temperatures and cultivation conditions (Passam, Lambropoulos, and Khan, 1997). Some genotypes withstand aging to a certain extent and remain viable for relatively longer periods under ambient conditions (Doijode, 1994). Such genotypes produce improved seed longevity in hybrids when used as parents.

Seed Collections

Peppers grow well in warm, humid climates. High temperature induces flowering, while excessive rainfall damages the crop, affecting the fruit set and causing fruit rotting. Peppers prefer deep, fertile, well-drained soils. Seeds are sown in nurseries on raised beds and subsequently transplanted to the field four to six weeks after sowing. In the field, adequate spacing, fertilization, weeding, and suitable plant protection measures must be followed to get a healthy crop. Though peppers are a self-pollinated crop, about 5 to

40 percent are cross-pollinated. Normally, bees, thrips, and ants cause cross-pollination. To prevent pollen contamination, an isolation distance of 400 m is adopted. Plants are inspected for vegetative, flowering, and fruiting characters during the flowering and fruiting stages, and off types are removed. Fruit matures four weeks after anthesis. Red ripe and disease-free fruits are plucked for seed purposes. Seeds are extracted manually or by machine, the latter giving better recovery at less expense (Karivaratharaju and Palaniswamy, 1984).

Fruit maturity. Mature bold seed maintains high seed quality during storage. Seed maturity coincides with that of fruit maturity and is expressed by change in fruit color. Early harvested fruits give poor-quality seeds with low viability and vigor. Fruits harvested at color breaker stage have high viability (Sayed and Essam, 1952). Metha and Ramakrishna (1988) reported that seeds extracted from fruits 48 days after anthesis have higher viability and vigor. Further, seeds extracted from second and third fruit pickings retain high viability after 12 months of ambient storage (Metha and Ramakrishna, 1986).

Seeds are fully capable of germination and impart tolerance to desiccation just before or at mass maturity. The maximum potential longevity occurs after 10 to 12 days of mass maturity (Demir and Ellis, 1992b). Seeds harvested at the ripe stage show higher germination and high seedling vigor (see Figure 34.3) (Doijode, 1988f), and can be stored longer under ambient conditions (see Figure 34.4) (Doijode, 1988c). Sanchez and colleagues (1993) noted that seeds extracted from mature green fruit did not germinate and required 14 days of storage for germination. Seeds from red fruits (50 days after anthesis) and overmature red fruits (60 days after anthesis) possessed higher germination capacity and greater dry weight. Seeds are gener-

FIGURE 34.3. Influence of Stage of Harvest on Seed Germination in Peppers

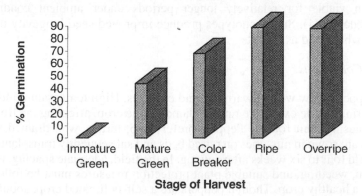

ally allowed to mature for a short period within the fruits after harvest for better seed quality.

Fruit storage. Dry pepper fruits can be stored for various periods, depending on the climatic and storage conditions. Fruit seeds stored up to six months exhibited higher germination (Radheshyam, Arora, and Tomer, 1996). Pepper fruits preserve well in cloth bags or paper bags, as these allow free exchange of gases and water vapor from the fruits. Seeds of these stored fruits gave 75 percent germination for three years, whereas seeds of fruits stored in sealed conditions in aluminum foil did not germinate (see Figure 34.5) (Doijode, 1996).

Seed position. Seed viability is influenced by the position of seeds inside the fruit. Seeds from the basal portion of the fruit have higher potential for viability and vigor as compared to those from the middle and tip (Doijode, 1991).

Seed Germination

Fresh seeds do not germinate readily and show the presence of dormancy. Seeds require six weeks of after-ripening to germinate (Randle and Honma, 1981). Several by-products, such as alcohol, aldehydes, ketones, esters, hydrocarbon, and furans, are released from seeds during after-ripening (Ingham et al., 1993). Pepper seeds germinate well at a constant temperature of 25 to 35°C and are inhibited below 20°C, which can be overcome by seed priming (PEG 6000 or potassium nitrate [KNO_3] + potassium orthophosphate [K_3PO_4]) (Giulianini et al., 1992). Seed germination is epigeal and takes about six to ten days to complete. Seeds of cultivars derived from other species, such as *Capsicum baccatum, C. chinense,* and *C. pubescens* do not germinate at a constant temperature and require an alternate temperature regime (Sato, Yazawa, and Namiki, 1982). Pepper seeds are sensitive to salinity, and germination is affected by high levels of salts, more than 4,000 ppm (Kaliappan and Rajagopal, 1970). Seed germination is rapid and higher when a little fruit pulp is retained along with the seed during drying. This aids better translocation of metabolites from pulp to seed at higher temperatures. Seed germination is enhanced by the use of plant stimulants such as atonik, and such treated seeds can also be stored for longer periods (Doijode, 1987a). Application of growth substances such as GA and IAA improves seed germination.

Storage of germinated seeds. Germinated seeds are more perishable and do not store long. However, when such seeds are placed in polyethylene bags (0.025 mm) with a vacuum seal or in nitrogen at 7°C they remain viable for 63 days (Ghate and Chinnan, 1987).

FIGURE 34.4. Stage of Harvest and Seed Longevity in Peppers

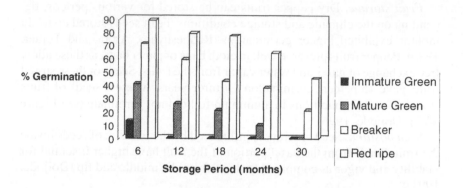

FIGURE 34.5. Fruit Storage and Seed Longevity in Peppers

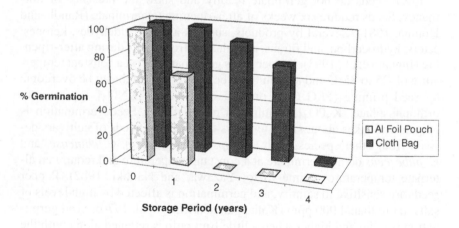

Storage Conditions

A high seed quality is essential for longer storage of seeds, and seed losses are minimal under ideal storage conditions. Initial seed quality, especially high vigor, contributes to longer storage life. Prevalence of high relative humidity and temperature shorten storage life. Chemical composition of seeds and atmospheric relative humidity determine the seed moisture content, thereby deciding seed longevity. The low level of seed moisture is maintained and protected from atmospheric humidity by suitable packaging. Sundstrom (1990) reported that 10 percent moisture is optimum for better storage of *Capsicum frutescens* seeds. In a study conducted by Popovska, Madenovski, and Mihajlovski (1981), pepper seeds (8 percent mc) were stored at 10.7 to 23.8°C and 45.7 to 85.7 percent RH for eight years in cloth bags, glass containers, polyethylene boxes, plastic bags, and tins. Among all, seed viability was highest in glassware-stored seeds, while in others, viability was lost gradually, and more rapidly in seeds stored in cloth bags. In a study by Fischer (1980), seeds remained viable for seven years when packed in sealed jars or plastic bags and stored at 5 to 10°C. However, seed lost viability rapidly under unsealed conditions. Under fluctuating temperatures from 2 to 26°C, seed viability decreased from 92 to 29 percent during five years of storage (Thakur et al., 1988). Ultra-low-temperature storage of seeds at –70°C did not improve the storability over that of –12°C (Woodstock et al., 1983). On aging, seeds became brown and lost vigor and viability (Doijode, unpublished data). Such seeds lose more electrolytes and soluble sugars on imbibitions through leaching (Doijode, 1988d).

Seed storage at low temperatures. Pepper seeds preserve well at low temperatures. Seed deterioration is less at 5°C storage (Passam, Lambropoulos, and Khan, 1997). High seed viability and vigor were maintained in seeds stored at 5 or –20°C after five years of storage, as compared to three years only when seeds were stored at ambient temperatures (Doijode, 1997b). In bell pepper, too, high seed viability was maintained at low (5°C) and subzero (20°C) temperatures for fifteen years, while under ambient conditions, bell pepper seeds remained viable for two years (see Figure 34.6) (Doijode, 1993). The loss of viability was associated with greater

FIGURE 34.6. Seed Viability and Vigor During Storage of Bell Pepper Seeds

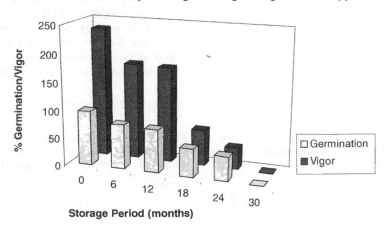

leaching of electrolytes, soluble sugars, and free amino acids from the seeds (Table 34.1) (Doijode, 1988e). Seeds with 4 percent moisture stored at 32.2°C maintained viability for at least three years (Anonymous, 1954). Further, seeds preserved viability for five years at 5°C and for ten years at –4°C in sealed containers (Barton, 1953). Furthermore, high seed viability was maintained in laminated aluminum foil pouches stored at 5 and –20°C for ten years. (Doijode, unpublished data).

Modified atmospheric storage of seeds. Seeds stored in partial vacuum under ambient conditions remain viable and vigorous for three years (Doijode, 1993). A higher concentration of carbon dioxide inhibits germination (Thornton, 1944) and preserves viability for five years under ambient conditions (Doijode, unpublished data).

Storage fungi. Pepper seeds associate with certain fungi under conditions of high seed moisture and high atmospheric RH (Mali and Joi, 1985). *Alternaria, Aspergillus, Chaetomium, Colletotrichum, Fusarium, Penicillium,* and *Rhizopus* species affect seed quality during storage, and these are effectively controlled with fungicidal applications (Mali, Joi, and Shinde, 1983). Fungicides, such as captan, carbendazim, copper oxychloride, and mancozeb are effective in maintaining good seed health. Sodium hypochloride and captan at 2.5 percent give effective control of storage fungi and high seed germination during storage (Fiume, 1994). The *Ocimum* oil is more effective than synthetic fungicides, and it does not show any adverse effect on seed germination or seedling growth during storage (Asthana et al., 1989).

Invigoration of Stored Seeds

Seed loses germinating capacity during storage. Such low-vigor seeds regain their quality to a certain extent when treated with growth substances or other physical stimuli. Seeds of cv. California Wonder showed improved germination after soaking in water for 72 h followed by gibberellic acid (50 ppm) application (Mostafa, Mohamedian, and Nassar, 1982). Likewise, seed quality improved with the application of potassium dihydrogen phosphate (3 percent) to four-year-old seeds (Solanki and Joshi, 1985), whereas

TABLE 34.1. Seed Viability and Leakages During Storage of Bell Pepper Seeds

Storage Temp. (°C)	Germination (%)	Dry Weight (mg)	Electrical Conductivity (mmhos)	Soluble Sugars (mg·g $^{-1}$)	Free Amino Acids (mg·g $^{-1}$)
Room	0	—	1,493	19.2	8.4
5	96	2.1	562	2.3	1.5
–20	97	2.6	674	4.5	2.3

repriming of primed seeds stored at 5°C caused reduction in seed germination (Jeong and Cho, 1996).

REFERENCES

Alekseev, R.V. 1976. The effect of low temperature on the germination of eggplant seed. *Ovoshchevodstra i Backchevodstva* No. 5:84-88.

Alekseev, R.V. and Prokhorov, I.A. 1982. Changes in the quality of tomato seeds extracted from fruits of different quality. *Biol. Osnovy Prom Tekhnol Ovoshchevoid Otkryt i Zakryt Grunta,* pp. 70-76.

Alvarado, A.D. and Bradford, K.J. 1987. Storage life and vigor of tomato *(Lycopersicon esculentum)* seeds following osmotic priming. *Acta. Hort.* No. 200:220.

Alvarado, A.D. and Bradford, K.J. 1988a. Priming and storage of tomato *(Lycopersicon lycopersicum)* seeds. I. Effects of storage temperature on germination rate and viability. *Seed Sci. Technol.* 16:601-612.

Alvarado, A.D. and Bradford, K.J. 1988b. Priming and storage of tomato *(Lycopersicon lycopersicum)* seeds. II. Influence of a second treatment after storage on germination and field emergence. *Seed Sci. Technol.* 16:613-623.

Anonymous. 1954. The preservation of viability and vigor in vegetable seeds. *Asgrow Monogr. New Haven* 2:32.

Argerich, C.A., Bradford, K.J., and Tarquis, A.M. 1989. The effect of priming and aging on resistance to deterioration of tomato seeds. *J. Exp. Bot.* 40:593-598.

Asthana, A., Dixit, K., Tripathi, N.N., and Dixit, S.N. 1989. Efficacy of ocimum oil against fungi attacking chilli seeds during storage. *Trop. Sci.* 29:15-20.

Balasubramanian, A. and Rangaswami, G. 1967. Studies on the influence of soil microorganism on seed germination and plant growth. *Mysore. J. Agric. Sci.* 1:1-6.

Baldo, N.B., Fonollera, V.C., Vallador, D.M., and Panilan, D.E. 1988. Influence of length and storage temp. on the seed viability and seedling growth of tomato. *CMU J. Sci. (The Philippines)* 1:48-58.

Barton, L.V. 1943. Effect of moisture fluctuations on the viability of seeds in storage. *Boyce Thompson Inst. Contrib.* No. 13:35-45.

Barton, L.V. 1953. Seed storage and viability. *Boyce Thompson Inst. Contrib.* No. 17:87-101.

Barton, L.V. and Garman, H.R. 1946. Effect of age and storage conditions of seeds on the yields of certain plants. *Boyce Thompson Inst. Contrib.* No. 14:243-255.

Beers, E.P. and Pill, W.G. 1986. Respiratory characterization of germinated tomato seeds stored in gels at low temperatures. *J. Am. Soc. Hort. Sci.* 111:918-921.

Berjak, P. and Villiers, T.A. 1973. Aging in plant embryo. II. Age induced damage and its repair during early germination. *New Phytol.* 71:135-144.

Bhatt, D.N.V. and Mazumdar, B.C. 1975. Effect of citrus juices on the relative inhibition of germination of tomato seeds. *Plant. Sci.* 7:37-40.

Bosewell, V.R., Toole, E.H., Toole, V.K., and Fisher, D.F. 1940. A study of rapid deterioration of vegetable seeds and methods for its prevention. *Tech. Bull. U.S. Dep. Agric.* 708:47.

Cavallaro, V., Mauromicale, G., and Vincenzo, G.Di. 1994. Effects of osmoconditioning on emergence characteristics of tomato *(Lycopersicon esculentum* Mill.). *Acta Hort.* No. 362:213-220.

Cucci, G., Caro, A.De., Ciciretti, L., and Leoni, B. 1994. Salinity and seed germination of some vegetable crops. *Acta Hort.* No. 362:305-309.

Demir, I., Ellialtioglu, S., and Tipirdamaz, R. 1994. The effect of different priming treatments on repairability of aged eggplant seeds. *Acta Hort.* No. 362:205-221.

Demir, I. and Ellis, R.H. 1992a. Changes in seed quality during seed development and maturation in tomato. *Seed Sci. Res.* 2:81-87.

Demir, I. and Ellis, R.H. 1992b. Development of pepper *(Capsicum annuum)* seed quality. *Ann. Appl. Biol.* 121:385-399.

Doijode, S.D. 1983. Studies on vigor and viability of seeds at different stages of fruit development in tomato. *Singapore J. Pri. Indus.* 11:106-109.

Doijode, 1987a. Effect of atonik on the longevity of chilli seeds under ambient conditions. *Prog. Hort.* 19:259-262.

Doijode, S.D. 1987b. Seed longevity in different tomato cultivars. *Prog. Hort.* 19:87-89.

Doijode, S.D. 1988a. Changes in seed quality on deterioration in tomato. *Prog. Hort.* 20:253-256.

Doijode, S.D. 1988b. Effect of storage environment on brinjal *(Solanum melongena)* seed viability. *Prog. Hort.* 20:292-293.

Doijode, S.D. 1988c. Seed storability as affected by different stages of fruit development in chilli. *Veg. Sci.* 15:15-20.

Doijode, S.D. 1988d. Solute leakage in relation to loss of seed viability in chilli cultivars. *Indian J. Pl. Physiol.* 31:285-287.

Doijode, S.D. 1988e. Studies on biochemical changes on deterioration in bell pepper seeds. *Capsicum Newsletter* 7:60-61.

Doijode, S.D. 1988f. Studies on vigor and viability of seeds as influenced by maturity in chilli. *Haryana J. Hort. Sci.* 17:94-96.

Doijode, S.D. 1991. Influence of seed position in fruit on seed viability and vigor during ambient storage of chilli fruits. *Capsicum Newsletter* 10:62-63.

Doijode, S.D. 1992. Genotypic differences for seed germinability longevity and leachates under accelerated aging in brinjal. *J. Mah. Agril. Universities* 18:107-109.

Doijode, S.D. 1993. Influence of partial vacuum on the storability of chilli *(Capsicum annuum)* seeds under ambient conditions. *Seed Res.* Spl. Vol. 1:322-326.

Doijode, S.D. 1994. Genotypic differences for seed vigor in chilli. *Prog. Hort.* 26:147-149.

Doijode, S.D. 1996. Effect of packaging and fruit storage on seed viability, vigor, and longevity in chilli *(Capsicum annuum L.)*. *Veg. Sci.* 23:36-41.

Doijode, S.D. 1997a. Effect of storage temperatures and packaging on longevity of tomato seeds. *Veg. Sci.* 24:70-72.

Doijode, S.D. 1997b. Influence of temperatures and containers on the storage of chilli seeds. *Indian J. Plant Genetic Resources* 10:163-166.

Doijode, S.D. 1998. Influence of seed packaging and storage conditions on seed viability and vigor in brinjal *(Solanum melongena L.)*. *Haryana J. Hort. Sci.* 27:33-36.

Doijode, S.D. and Raturi, G.B. 1990. Effect of hydration-dehydration on the storability of tomato and radish seeds. *Indian J. Pl. Physiol.* 33:172-174.

Eguchi, T. and Yamada, H. 1958. Studies on the effect of maturity on longevity in vegetable seeds. *Natl. Inst. Agri. Sci. Bull. Ser. E Hort.* 7:145-165.

Fischer, I. 1980. Studies on storage conditions affecting the germinating capacity of capsicum seeds. *Zoldsegtermeszteri Kutato Intezet Bull.* 14:13-20.

Fiume, F. 1994. Seed disinfection of pepper (*Capsicum annuum* L.) from *Alternaria nees. Acta Hort.* 362:311-318.

Geetharani, P., Ponnuswamy, A.S., and Raju, T.V.K. 1996. The influence of midstorage seed treatment on yield and quality of resultant seeds in tomato and brinjal. *Madras Agril. J.* 83:737-738.

Ghate, S.R. and Chinnan, M.S. 1987. Storage of germinated tomato and pepper seeds. *J. Am. Soc. Hort. Sci.*112:645-651.

Giulianini, D., Nuvoli, S., Pardossi, A., and Tognoni, F. 1992. Pregermination treatment of tomato and pepper seeds. *Colture Protette* 21:73-79.

Harrington, J.F. 1960. Drying, storing, and packaging seeds to maintain germination. *Seedsmen Digest* 1:16-18.

Harrington, J.F. and Setyati-Harjadi, S. 1966. The longevity of vegetable seeds under high temperature and high humidity. *Proc. 17th Inter. Hort. Congr. Md.* 1:627.

Ingham, B.H., Hsieh, T.C.Y., Sundstrom, F.J., and Cohn, M.A. 1993. Volatile compound released during dry after-ripening of tabasco pepper seeds. *J. Agril. Food Chem.* 41:951-954.

James, E., Bass, L.N., and Clark, D.C. 1964. Longevity of vegetable seeds stored 15 to 30 years at Cheyenne, Wyoming. *Proc. Am. Soc. Hort. Sci.* 84:527-533.

Jayaraj, T., Vadivelu, K.K., Dharmalingam, C., Vijayakumar, A., and Irulappan, I. 1988. Effect of seed treatments and containers on vegetable seed storage under different agroclimatic conditions. *South Indian Hort.* 36:183-187.

Jeong, Y.O. and Cho, J.L. 1996. Effect of storage temperature and repriming of primed pepper (*Capsicum annuum* L.) seeds on the germinability. *J. Korean Soc. Hort. Sci.* 37:201-205.

Kaliappan, R. and Rajagopal, A. 1970. Effect of salinity on the south Indian field crops, germination and early vigor of chillies *(Capsicum annuum). Madras Agric. J.* 57:231-233.

Karivaratharaju, T.V. and Palaniswamy, V. 1984. Effect of seed extraction methods on seed quality in chillies. *South Indian Hort.* 32:243-244.

Kartapradja, R. 1988. The effect of seed size on tomato yield. *Bull. Penel. Hort.* 16:34-36.

Khan, B.A. and Stoffella, P.J. 1996. No evidence of adverse effects on germination, emergence and fruit yield due to space-exposure of tomato seed. *J. Am. Soc. Hort. Sci.* 121:414-418.

Kretschmer, M. 1981. Storage of seeds of some unusual vegetables. *Deut. Gartenbauwissenschaft* 36:1994-1997.

Krishnasamy, V. and Rangarajpalaniappan, M.S.M.G. 1990. Studies on seed dormancy in brinjal var. Annamalai. *South Indian Hort.* 38:42-44.

Kurdina, V.N. 1966. Changes in the sowing quality of seeds of vegetable crops during storage. *Izv. timirjazev Sel hoz Akad* No. 5:135-144.

Liu,Y.Q., Bino, R.J., Burg, W.J., Groot, S.P.C., and Hilhorst, H.W.M. 1996. Effect of osmotic priming on dormancy and storability of tomato *(Lycopersicon esculentum)* seeds. *Seed Sci. Res.* 6:49-55.

Mali, J.B., Joi, M.B., and Shinde, P.A. 1983. Fungi associated with chilli seeds. *J. Mah. Agril. Universities* 8:69-71.

Mali, J.B. and Joi, M.B. 1985. Control of seed mycoflora of chilli *(Capsicum annum)* with fumigants. *Curr. Res. Rep.* 1:8-10.

Mancinelli, A.L., Borthwick, H.A., and Hendricks, S.B. 1966. Phytochrome action in tomato seeds production. *Bot. Gaz.* 127:1-5.

Mayee, C.D. 1977. Storage of seed for pragmatic control of a virus causing mosaic diseases of brinjal. *Seed Sci. Technol.* 5:555-558.

Metha, V.A. and Ramakrishna, V. 1986. Effect of different pickings on the storability of chilli seeds. *Madras Agric. J.* 73:661-667.

Metha, V.A. and Ramakrishna, V. 1988. Studies on fruit and seed development in chilli. *Madras Agric. J.* 75:334-339.

Mitra, R. and Basu, R.N. 1979. Seed treatment for viability, vigor and productivity of tomato. *Scientia. Hort.* 11:365-369.

Mobayen, R.G. 1980. Germination and emergence of citrus and tomato seeds in relation to temperature. *J. Hort. Sci.* 55:291-297.

Mohamed, E.I., Lester, R.N., and Mumford, P.M. 1988. Viability of eggplant seeds. *Trop. Agril.* 65:279-280.

Mostafa, H.A.M., Mohamedian, S.A., and Nassar, S.M. 1982. A study on improving germination of old seeds of sweet pepper *(Capsicum annuum)*. *Res. Bull. Fac. Agr. Ain Shams University* No. 1808:17.

Naik, L.B., Prabhakar, M., Hebbar, S.S., and Doijode, S.D. 1995. Influence of order of fruit formation on seed contents and its quality in brinjal *(Solanum melongena* L.). *Seed Res.* 23:71-74.

Nakamura, S., Sato, T., and Mine, T. 1972. Storage of vegetable seeds dressed with fungicidal dusts. *Proc. Inter. Seed Test. Assoc.* 37:961-968.

Nelson, S.O., Nutile, G.E., and Stetson, L.E. 1970. Effect of radio frequency electrical treatment on germination of vegetable seeds. *J. Am. Soc. Hort. Sci.* 95:359-366.

Odell, G.B. and Cantliffe, D.J. 1987. Seed priming procedures and the effect of subsequent storage on the germination of fresh market tomato seeds. *Proc. Fl. Stat. Hort. Soc.* 99:303-306.

Owen, P.L. and Pill, W.G. 1994. Germination of osmotically primed asparagus and tomato seeds after storage up to three months. *J. Am. Soc. Hort. Sci.* 119:636-641.

Palaniappan, E., Muthukrishnan, C.R., and Irulappan, I. 1981. Growth analysis for seed germination, seedling weight and yield of fruit in tomato inbreds and hybrids. *Madras Agric. J.* 68:366-372.

Pandita, V.K. and Randhawa, K.S. 1995. Influence of seed size grading on seed quality of some tomato cultivars. *Seed Res.* 23:31-33.

Passam, H.C., Lambropoulos, E., and Khan, E.M. 1997. Pepper seed longevity following production under high ambient temperatures. *Seed Sci. Technol.* 25:177-185.

Petrikova, K. 1989. Results of the effect of four temperatures on tomato seed swelling. *Acta University Agric. Fac. Hort.* 4:47-51.

Petrov, H. and Dojkov, M. 1970. Influence of some factors on seed quality and germinability in eggplants. *Gradinarstvo.* 128:29-31.

Pijlen, J.G., Groot, S.P.C., Kraak, H.L., Bergervoet, J.H.W., and Bino, R.J. 1996. Effect of prestorage hydration treatments on germination performance, moisture content, DNA synthesis and controlled deterioration tolerance of tomato (*Lycopersicon esculentum* Mill.) seeds. *Seed Sci. Res.* 6:57-63.

Pill, W.G. and Fieldhouse, D.J. 1982. Emergence of pregerminated tomato seed stored in gels up to twenty days at low temperatures. *J. Am. Soc. Hort. Sci.* 107:722-725.

Pollock, B.M. and Roos, E.E. 1972. Seed and seedling vigor. In *Seed biology,* Volume I, Kozolowski, T.T. (Ed.). New York: Academic Press, pp. 313-387.

Ponappa, K.M. and Mazumdar, B.C. 1976. Germination behavior of tomato seeds grown in juice extracts of emblica fruits. *Prog. Hort.* 8:19-24.

Popovska, P., 1964. A contribution to the knowledge of the effect of seed age on the total germination and germinating power in seeds of peppers and tomato. *Ann. Fac. Agric. Syloic. Skopje Agric.* 17:165-177.

Popovska, P., Madenovski, Lj.T., and Mihajlovski, M. 1981. The influence of packing over germination of pepper and tomato seeds. *Acta Hort.* 111:281-290.

Quagliotti, L. and Rota, A. 1986. Germination trials of eggplant seeds (*Solanum melongena* L.). *Capsicum Newsletter* No. 5:72-73.

Radheshyam, Arora, S.K., and Tomer, R.P.S. 1996. Effect of seed extraction interval on seed storability of chilli *(Capsicum annuum)* cultivars. *J. Res.* 286:183-186.

Randle, W.M. and Honma, S. 1981. Dormancy in peppers. *Scientia Hort.* 14:19-25.

Reddy, V.S. and Reddy, B.M. 1994. Effect of seed protectants on storability of eggplant (*Solanum melongena* L.) seed. *Seed Res.* 22:181-183.

Sanchez, V.M., Sundstrom, F.J., McClure, G.N., and Lang, N.S. 1993. Fruit maturity, storage and post-harvest maturation treatment affects bell pepper (*Capsicum annuum* L.) seed quality. *Scientia Hort.* 54:191-201.

Santos, D.Dos. and Yamaguchi, M. 1979. Seed sprouting in tomato fruits. *Scientia Hort.* 11:131-139.

Sato, T., Yazawa, S., and Namiki, T. 1982. Temperature requirements for capsicum seed germination. *Rep. Kyoto Pre. University Agr.* No. 34:21-27.

Sayed, M. Saakr and Essam Mahmoud, El-Din. 1952. Viability of seeds harvested from fruits at different stages of maturity. *Proc. Am. Soc. Hort. Sci.* 60:327-329.

Selvaraj, J.A. 1988. Effect of density grading on seed quality attributes in brinjal. *South Indian Hort.* 36:32-35.

Shuck, A.L. 1936. The germination of secondary dormancy in tomato seeds and their formation. *Proc. Inter. Seed Test. Assoc.* 8:136-158.

Singh, G., Singh, H., and Dhillon, T.S. 1985. Some aspects of seed extraction in tomato. *Seed Res.* 13:67-72.

Singh, R. and Lal, G. 1990. Effect of varieties and harvesting stages on yield and quality of tomato seeds *(Lycopersicon esculentum). Inter. Conf. Seed. Sci. Technol.* Abstr. No. 1.37, p. 19.

Solanki, S.S. and Joshi, R.P. 1985. Effect of different chemicals on invigoration in seed germination of cucumber *(Cucumis sativus)* and capsicum *(Capsicum annuum). Prog. Hort.* 17:122-124.

Spaldon, E. and Pevna, V. 1966. A contribution to pepper seed germination energy and germination capacity studies. *Acta Fytotech Nitra.* 14:5-14.

Sundstrom, F.J. 1990. Seed moisture influences on tabasco pepper seed viability, vigor, and dormancy during storage. *Seed Sci. Technol.* 18:179-185.

Sunil, G.D.J.L. 1991. Storage of pregerminated vegetable seeds. *Laguna Coll. Tech. Bull.,* p. 123.

Suryanarayana, V. and Rao, V.K. 1984. Effect of growth regulators on seed germination in okra, tomato, and brinjal. *Andhra Agric. J.* 31:220-224.

Suzuki, Y. and Takahashi, N. 1968. Effect of after-ripening and gibberellic acid on the thermo induction of seed germination in *Solanum melongena. Plant Cell Physiol.* 9:653-660.

Thakur, P.C., Joshi, S., Verma, T.S., and Kapoor, K.S. 1988. Effect of storage period on germination of sweet pepper seeds. *Capsicum Newsletter* 7:58-59.

Thornton, N.C. 1944. Carbon dioxide storage. XII. Germination of seeds in the presence of carbon dioxide. *Boyce Thompson Inst. Contrib.* 13:355-360.

Thulasidas, G., Selvaraj, J.A., and Thangaraj, M. 1977. Seed storage in brinjal *(Solanum melongena). Madras Agric. J.* 64:646-649.

Tripathi, R.S. and Srivastava, P.P. 1970. Effect of aqueous plant extracts on the seed germination of *Lycopersicon esculentum* Mill. *Sci. Cult.* 36:59-60.

Varier, A. and Agrawal, P.K. 1989. Long term storage of certain vegetable seeds under ambient and reduced moisture conditions. *Seed Res.* 17:153-158.

Vos, C.H.R.De., Kraak, H.L., and Bino, R.J. 1994. Aging of tomato seeds involves glutathione oxidation. *Physiol. Plant.* 92:131-139.

Went, F.W. 1961. Problems in seed viability and germination. *Proc. Inter. Seed Test. Assoc.* 26:674-685.

Winden, C.M.M. and Bekendam, J. 1975. Germination of eggplant seed. *Groentenen Fruit* 31:729.

Woodstock, L.W., Maxon, S., Faul, K., and Bass, L.N. 1983. Use of freeze drying and acetone impregnation with natural and synthetic antioxidants to improve storability of onion, pepper, and parsley seeds. *J. Am. Soc. Hort. Sci.* 108:692-696.

Yaovalak, T. 1987. Effect of seed development, color and size of fruits on seed quality of tomato cv. Sida. *KU Bangkok* p. 42.

35

Okra: *Abelmoschus esculentus* (L.) Moench

Introduction

Okra is commonly called gombo, gumbo, and lady's finger. It is cultivated for its edible pods. The immature pods are boiled, fried, or cooked as a vegetable. Tender fruits are also canned or dehydrated. Plant extract is used for clarifying sugarcane juice in jaggary preparation. Sometimes fiber is extracted from the stem and used in the textile industry. Pods are rich in vitamins and minerals such as calcium and potassium. Seeds are rich in oil and contain about 20 percent edible oil.

Origin and Distribution

Large genetic diversity in okra is found in eastern Africa. Okra is presumed to have originated in Ethiopia, and the wild forms are distributed in Ethiopia and Sudan. Okra is grown in the tropics, subtropics, and, during summer, in temperate regions. It is cultivated in China, Egypt, India, Indonesia, Malaysia, Nigeria, the Philippines, Puerto Rico, Sudan, and Trinidad.

Morphology

Okra belongs to the mallow family Malvaceae, and its chromosome number is 2n = 72 and 144. Plant is an erect, woody, dicotyledonous annual with well-developed taproots. Shoots grow up to 1 to 2 m heights. Stems are green or tinged red; leaves are broad, alternate, hairy, three to five lobed, and cordate. Flowers are single, perfect, axillary, yellow, bell shaped, with epicalyx, five sepals, and five petals; stamens are many, and the ovary is superior. Fruit is a capsule that dehisces longitudinally on drying. It is light to dark green or purple, round or ridged. Seeds are round, medium, and dark green, gray, or black, and 30 to 80 seeds are present in one pod. One thousand seeds weigh about 50 g.

Seed Storage

Okra is commercially cultivated through the seeds. Its seeds show orthodox storage behavior, in which the longevity improves with desiccation and storage at chilling temperatures.

Seed Collections

Okra grows best under warm conditions. The optimum temperature for germination is 25 to 35°C, and no germination occurs below 16°C. Rains during fruit development and maturity affect seed quality (Singh and Singh, 1988). It needs fertile, well-drained soil. Seeds are directly sown on the ridges in the field. Regular weeding, irrigation, fertilizer application, and plant protection measures are followed. Plants are self-pollinated, but about 20 percent cross-pollination occurs by insects. Isolation distance of 400 m is maintained to avoid pollen contamination. Plants are inspected for various morphological characteristics during the flowering and fruiting stages, and off-type plants are removed. Flowering starts 35 to 60 days after planting (see Figure 35.1). Fruits turn from green to gray or brown and become hard on ripening. Seeds obtained from basal pods are heavier and of good quality. Seeds separated from dried pods are cleaned, graded, and dried before storage.

Fruit position. Seed viability and vigor are affected by the fruit's nodal position. Seed quality is better in seeds collected from fruits borne at the third to sixth nodes (Prabhakar et al., 1985). Seeds from fruits borne up to the seventh node showed higher germination and seedling vigor. Seed longevity decreases with an increase in fruit nodal position (see Table 35.1) (Doijode, 1999).

Fruit maturity. Fruit fiber content increases with advancing maturity. Fruit becomes hard and turns gray or brown. Seed moisture is 85 to 90 percent at earlier stages of seed development, and it decreases to 19 to 24 percent on seed maturity. Reduction of seed moisture and greater dry weight accumulation are rapid at 30 to 35 days of anthesis. Seeds up to ten days of development do not germinate; thereafter, germination increases. Seed attains maximum viability and vigor between 30 to 35 days after fruit development. When a crack appears on the fruits, they are ready for harvest (Velumani and Ramaswamy, 1976). Further, delay in fruit harvesting affects seed quality (Kanwar and Saimbhi, 1987). Demir (1994) observed that maximum seed dry weight occurs at 31 days after anthesis, with a moisture content of 71 percent on a dry weight basis, but high seed quality in terms of germination was recorded 52 days after anthesis, with seed moisture content at 12 percent and fully matured seeds.

FIGURE 35.1. Okra Seed Production Plot

TABLE 35.1. Nodal Position of Fruit and Seed Germination (Percent) During Ambient Storage of Okra Seeds

Nodal Position	Initial Germi- nation	1st Year of Storage	2nd Year of Storage	3rd Year of Storage	4th Year of Storage	5th Year of Storage
3	97	98	95	93	83	29
4	97	100	99	97	87	28
5	98	97	96	97	83	30
6	98	99	93	95	83	22
7	94	98	95	97	82	18
8	94	93	88	95	72	12
9	94	92	83 ˟	91	72	11
10	95	90	80	75	65	0

Seed Germination

Seed germination is epigeal and takes five to seven days to complete. However, it is delayed due to the presence of a hard seed coat. Seed germination is better at a constant temperature of 25 to 35°C or at alternate temperatures of 20/30°C for 16/8 h, respectively.

Seed dormancy. The seed coat is somewhat hard, and the embryo develops slowly during germination. High-moisture seeds tend to show little or no dormancy. Reduction of moisture to 4 to 6 percent causes hard-seededness, which leads to slow uptake of water. To overcome this, seeds are scarified for higher germination (Medina, Medina, and Shimoya, 1972). Likewise, seed germination at alternate temperatures (20/30°C for 16/8 h) removes the dormancy. Further, exposing seeds to 40°C for four days increases the water uptake and total germination (Nada, Lotito, and Quagliotti, 1994). Use of a growth regulator such as gibberellic acid (50 ppm) improves the germination percentage and total yield (Pawar, Joshi, and Mahakal, 1977). Seed exposure to physical stimuli such as a 40 megahertz (MHz) radio frequency electric field reduced the number of hard seeds and gave higher germination (Nelson, Nutile, and Stetson, 1970). Dutta (1972) recorded the stimulating effects of gamma rays on okra seed germination up to 60 kr dosages. Beyond this dose, germination percentage was reduced.

Seed color. Okra seed color ranges from green to gray and black. Black seeds show poor seed quality. Seed germination and seedling vigor are rapid and higher in green than in black seeds (Singh and Gill, 1983; Baruah and Paul, 1996).

Seed size. Seed size varies in okra fruit. Seeds from the peduncle end are larger, heavier, and bolder than those from apical end. Large seeds show better germination than small seeds after 12 months of ambient storage (27°C and 67 percent RH) (Coelho et al., 1984).

Storage Conditions

Okra seeds preserve well under ambient conditions. High storage temperature and high relative humidity reduce the storage life of seeds. Therefore, dry, cool conditions are to be created for longer storage of seeds.

Seed moisture. Low seed moisture, preferably 5 to 7 percent, is ideal for seed storage. Agrawal (1980) stored okra seeds (10 percent mc) for 37 months in ambient conditions. Moisture content of 13.6 percent or less is ideal for maintaining high seed viability for short periods, while 8 percent moisture maintains viability for seven years at room temperature. Further high-moisture seeds (19 and 25 percent) failed to germinate after one year of storage at room temperature, and after five years in cold storage. Reduction of seed moisture slows down the deterioration process. Seeds with 12 percent or less moisture stored in cold storage retained 90 percent germination capacity for 11 years (Martin, Seenn, and Crawford, 1960).

Storage temperature. Low storage temperature is congenial for long-term storage. However, high-moisture seeds are not suitable for sealed storage at low temperature. High seed viability and vigor are maintained for ten years at 5 and –20°C, as compared to three years at room temperatures (see

Figure 35.2) (Doijode, unpublished data). Seed viability rapidly decreases at higher storage temperatures, which promote the leaching of electrolytes and solute sugar from the seeds (Doijode, 1986).

Seed packaging. Packages protect seeds from various extraneous factors, such as high relative humidity, insects, and fungi. Seeds are commonly stored in polyethylene bags, cloth bags, glass containers, laminated aluminum foil pouches, or aluminum cans, depending upon storage period required, cost of seed material, and storage conditions. Cloth and paper bags are generally suitable for open storage for shorter periods, and moisture-proof packages are used for longer storage at low or subzero temperatures. Polyethylene bags are effective in maintaining high viability at 5°C, and laminated aluminum foil pouches at –20°C (Doijode, 1997). Seeds stored in glass containers at room temperature and in cold storage retained germination capacity for 29 months. Even seeds stored in polyethylene bags show similar effects, unlike those stored in paper bags, which showed low germination percentage (Silva et al., 1976).

Air. High levels of oxygen and moisture promote metabolism and shorten the storage life of seeds. High-moisture okra seeds stored in a high level of oxygen and a low level of carbon dioxide showed poor germination, while high carbon dioxide and low oxygen levels tend to maintain seed viability (Martin, Seenn, and Crawford, 1960).

Storage fungi. Storage fungi affect seed quality and give low germination. Normally, fungi grow at high levels of seed moisture. Okra seeds associate with fungi such as *Actinomucor* sp., *Alternaria tenuis, Aspergillus*

FIGURE 35.2. Influence of Temperatures on Seed Viability and Longevity in Okra

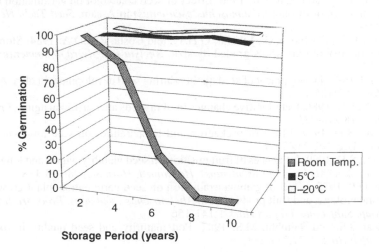

flavus, A. niger, A. sulphureus, A. terreus, Cephalosporium curtipes, Cehaleospora irregularis, Fusarium moniliforme, Gonatorrhodiella sp., and *Rhizopus nigricans* (Saxena, Kumari, and Karan, 1982). High seed moisture encourages the growth of *Botrytis* spp. at room temperatures of 21.1 to 37.8°C and *Penicillium* spp. both at room temperatures and in cold storage (1.7 to 4.4°C), thereby reducing germination. High seed germination can be maintained for 25 months by treating seeds with captan (2 grams per kilogram [g·kg⁻¹]) (Palanisamy and Vanangamudi, 1987).

Invigoration of Stored Seeds

Seed deterioration causes a decline in germination and vigor and is influenced by the chronological age of the seeds (Agrawal and Sinha, 1980). Germination, vegetative growth, reproductive growth, and yield can be increased by the application of gibberellic acid (50 ppm) to aged seeds (Kumar et al., 1996) and by the application of indoleacetic acid (100 ppm) to one-year-old seeds (Suryanarayana and Rao, 1984).

REFERENCES

Agrawal, P.K. 1980. Relative storability of seeds of ten species under ambient conditions. *Seed Res.* 8:94-99.

Agrawal, P.K. and Sinha, S.K. 1980. Response of okra seeds *(Abelmoschus esculentus)* to different chronological ages during accelerated aging and storage. *Seed Res.* 8:64-70.

Baruah, G.K.S. and Paul, S.R. 1996. Effect of seed coat color on germination and seedling vigor of okra *(Abelmoschus esculentus)* in Assam. *Seed Tech. News.* 26:3-4.

Coelho, R.C., Coelho, R.G., Liberal, O.H.T., and Costa, R.A.D.A. 1984. Storage and quality of okra seeds graded by size. *Revista Brazileira de Sementes* 6: 17-27.

Demir, I. 1994. Development of seed quality during seed development in okra. *Acta Hort.* 362:125-131.

Doijode, S.D. 1986. Deteriorative changes in okra seeds after artificial aging. *Prog. Hort.* 18:218-221.

Doijode, S.D. 1997. Effect of packaging and temperatures on seed longevity in okra. *Veg. Sci.* 24:67-69.

Doijode, S.D. 1999. Influences of fruit position on seed viability, vigor, and longevity in okra *(Abelmoschus esculentus).* *Haryana J. Hort. Sci.* 28:135-138.

Dutta, O.P. 1972. Effect of gamma irradiation on seed germination, plant growth, floral biology, and fruit production in *Abelmoschus esculentus.* *Proc. 3rd Inter. Symp. Subtropics Tropics Hort.* 1:141-156.

Kanwar, J.S. and Saimbhi, M.S. 1987. Pod maturity and seed quality in okra. *Punjab Hort. J.* 27:234-238.

Kumar, S., Singh, P., Katiyar, R.P., Vaish, C.P., and Khan, A.A. 1996. Beneficial effect of some plant growth regulators on aged seeds of okra under field conditions. *Seed Res.* 24:11-14.

Martin, J.A., Seenn, T.L., and Crawford, J.H. 1960. Response of okra seed to moisture content and storage temp. *Proc. Am. Soc. Hort. Sci.* 75:490-494.

Medina, P.V.L., Medina, R.M.T., and Shimoya, C. 1972. Okra seed coat anatomy and the use of chemicals to hasten germination. *Revista Ceres* 19:385-394.

Nada, E., Lotito, S., and Quagliotti, L. 1994. Seed treatments against dormancy in okra *Abelmoschus esculentus. Acta Hort.* No. 362:133-140.

Nelson, S.O., Nutile, G.E., and Stetson, L.E. 1970. Effect of radio frequency electrical treatment on germination of vegetable seeds. *J. Am. Soc. Hort. Sci.* 95:359-366.

Palanisamy, V. and Vanangamudi, K. 1987. Viability of okra seeds in storage. *Seed Res.* 15:221-222.

Pawar, P.R., Joshi, A.T., and Mahakal, K.G. 1977. Effect of seed treatment with plant growth regulators on germination, growth and yield of okra. *J. Mah. Agril. Universities* 2:26-29.

Prabhakar, B.S., Hegde D.M., Srinivas K., and Doijode, S.D. 1985. Seed quality and productivity of okra in relation to nodal position of pod. *South Indian Hort.* 33:115-117.

Saxena, N., Kumari, V., and Karan, D. 1982. Mycoflora associated with seeds of okra *(Abelmoschus esculentus). Seed Res.* 10:175-176.

Silva, R.F., Silva, J.F., Viggiano, J., Couto, F.A., and Conde, A.R. 1976. The effect of seed moisture content, type of containers and storage conditions on the germination of okra seeds. *Revista Ceres* 23:77-82.

Singh, G. and Singh, H. 1988. Effect of simulated rains on seed quality of okra cultivars. *Seed Res.* 16:226-228.

Singh, H. and Gill, S.S. 1983. Effect of seed coat color on seed germination of okra *(Abelmoschus esculentus* L.). *Seed Res.* 11:20-23.

Suryanarayana, V. and Rao, V.K. 1984. Effect of growth regulators on seed germination in okra, tomato and brinjal. *Andhra Agric. J.* 31:220-224.

Velumani, N.P. and Ramaswamy, K.R. 1976. Studies on seed germination in bhendi *(Abelmoschus esculentus). Madras Agric. J.* 66:75-83.

36

Bulb Crops

ONION: *Allium cepa* L.

Introduction

Onion is a popular bulb vegetable crop in temperate, tropical, and subtropical regions. Both immature and mature bulbs are eaten raw, cooked, or used in the preparation of different vegetable dishes. The green leaves are also used in salads, and cooked as a vegetable. Onion contains high amounts of carbohydrates, calcium, phosphorus, vitamins A and B, and the volatile substance allyl propyl disulphide. Onion has medicinal value in that it possesses antibacterial properties.

Origin and Distribution

Onion originated in Central Asia comprising Afghanistan, Iran, Tajikstan, Turkey, and West Pakistan. It is cultivated in Argentina, Brazil, China, Colombia, Egypt, Ghana, India, Indonesia, Iran, Italy, Japan, Korea, Malaysia, Mexico, Morocco, Myanmar, the Netherlands, Nigeria, Pakistan, the Philippines, Poland, Russia, South Africa, Spain, Turkey, and the United States.

Morphology

Onion belongs to the family Alliaceae and has chromosome number of $2n = 16$. It is a monocotyledonous biennial herb but is cultivated as an annual. Stems are short with poorly developed root systems, and adventitious roots emerge from the base of the stems. Leaves are alternate, hollow, and arranged in a circle, forming a bulb. Bulbs are flat, globular, or cylindrical in shape. Bulb color varies, being either white, yellow, brown, pink, red, or green. Inflorescence is a terminal umbel consisting of 50 to 2,000 tiny flowers. Number of seed stalks varies from 1 to 20, depending upon bulb size, cultivars, and growing conditions. Flowers are white, with six stamens and a protoandrous superior ovary, and trilocular. Fruit is a capsule that splits longitudinally and contains one to two seeds. Seeds are black, shiny, and wrinkled. Embryo is crescent shaped or curved and embedded in the endosperm. One thousand seeds weigh about 3.5 g.

Seed Storage

Onion is propagated by seeds. Seeds are short-lived under ambient conditions. Seeds are also used for evolving new varieties in breeding programs, and for long-term conservation of genetic diversity. Onion seeds exhibit orthodox storage behavior; they can withstand loss of moisture and have extended longevity under low-temperature storage conditions. Seed longevity of several crop species was studied under different atmospheric conditions; and onion seeds were found to be the most short-lived (Duvel, 1905; Pritchard, 1933). Ellis and Roberts (1977) also reported that onion seeds lost viability more rapidly than seeds of other crops. Onion seeds preserve viability for 6 to 12 months under ambient conditions (Agrawal, 1980; Doijode, 1989).

Seed longevity is primarily governed by genetic means. However, it is also regulated by storage conditions. It varies in different genotypes, species, and cultivars. Good-storer cultivars withstand relatively higher temperatures and relative humidity. Doijode (1990d) identified certain good and poor-storer cultivars. Poor-storer types showed greater cellular damage during storage, as evidenced by excessive leakage of electrolytes and soluble sugars. These cultivars are useful in studying inheritance of seed storability and in hybridizing for better storage quality.

Location of Seed Storage

Extreme climatic conditions affect seed longevity. Areas where dry and cool climates prevail during most of the year are suitable for safe seed storage. Agrawal (1976) identified the places suited for short-term seed storage based on prevailing temperatures and relative humidity during a year. Harrington (1960b) suggested that the combination of temperature (°F) and relative humidity (percent) should not exceed 100 for better storage.

Seed Collections

Onion is a cool-season crop that requires low temperatures for bulb formation and flowering. It withstands frost. Plants grown under low temperatures (15 to 16°C) yielded more seeds than those grown under high temperatures (22 to 23°C) (Gray and Steckel, 1984). Heavy rainfall areas are not suited for cultivation. Deep, fertile, sandy loam soil with good drainage is ideal for onion cultivation. Onion has a shallow root system and needs frequent irrigation.

Onion is propagated by seeds. Seeds are sown in a nursery, and four-week-old seedlings are transplanted to the field. Soil sterilization is useful, especially in nursery bed preparation, for reducing pink root rot disease and for higher seedling vigor (Katan, Abramski, and Levi, 1975). Seeds are produced by the bulb-to-seed or seed-to-seed method. In the former method, mature healthy bulbs are planted for seed production. Bulbs are stored for four to five months before planting for curing. Bulbs stored at 10°C for 90 days flowered earlier, produced more flowers, and had the highest seed yield (Behairy and El-Habbasha, 1979; Hesse, Vest, and Honma, 1979). Likewise, bulbs stored for 12 weeks at 10°C resulted in higher yield (Hesse, Vest, and Honma, 1979).

Onion yield increases with increased seed size (Gamiely, Smitte, and Mills, 1990). Onion is a cross-pollinated crop, and pollination is mainly brought about by bees (see Figure 36.1). An isolation distance of 1,600 m is used to maintain genetic purity. Plants are adequately fertilized, and protected against pests and diseases. Plants with double bulbs and thick necks are removed to maintain varietal quality. Heads are collected when fruits open and expose the black seeds (see Figure 36.2). Later they are cured for two to three days and then threshed; seeds are separated, cleaned, dried, and suitably packed for storage.

Seed maturity. Bold, uniform, healthy, and high-vigor seeds are best suited for storage. Harrington (1972) opined that seeds attained maximum dry

FIGURE 36.1. Onion Plants Protected from Insects

FIGURE 36.2. Onion Heads Ready for Harvest

weight during maturity. These mature seeds store longer than immature seeds. Sandhu, Nandpuri, and Thakur (1972) observed that seeds harvested one week prior to full maturity gave normal percentage of germination, while germination decreased in seeds harvested two weeks early. Further, a delay in harvesting affects seed quality by risking rain damage or seed shattering.

Seed drying. Seeds are capable of exchanging moisture with the atmosphere. The process is faster in seeds rich in storage proteins, and it is slower in onion seeds. Seeds absorb moisture from the atmosphere when the vapor pressure is lower in seeds, and they lose moisture when the vapor pressure is lower in the atmosphere. When the two vapor pressures are equal, no movement of moisture vapor occurs, and the system is in equilibrium. Seeds dry on the movement of water vapor out of the seeds into the atmosphere by creating a moisture gradient. Seed structure will be damaged at a higher rate of evaporation. Thus, seeds are preferably dried at low temperature and low humidity. Drying seeds with dehumidified air at 25 to 40°C did not affect the germination percentage (Pohler, 1984). Seeds are dried by various means, such as natural drying; sun drying; heated, unheated, and dehumidified air drying; drying with desiccants; and vacuum and freeze-drying. Each has advantages and disadvantages. Seeds are protected against heat damage during drying. Seeds dried in scorching heat and sun showed reduced germination. Copeland (1976) noted that drying too rapidly at a high temperature was injurious to seed life. The reduction of seed moisture from 10.0 to 6.5 percent is more beneficial for preserving high viability than nondrying of seeds. Drying with silica gel is optimum for maintaining high seed viability for a longer period (Doijode, 1990b). Earlier, Barton (1935) believed that onion seeds should be dried to one-third to one-half of their original moisture content for satisfactory storage at room temperatures.

There are different opinions about ultralow-mc drying of seeds: It damages the cell structure and contents, thus affecting viability and seed storability, and also it is unsuitable for storage of large quantities of seeds. Freeze-drying for one day increases the seed longevity (Woodstock, Simkin, and Schroeder, 1976). Freeze-dried seeds showed 61 percent germination after one year of storage at 50°C in sealed containers (Woodstock and Schroeder, 1975). Further excessive drying is to be avoided, as water in the protective monolayer should not be removed (Woodstock, 1975). Furthermore, freeze-dried seeds (4.1 percent mc) stored in desiccators over zeolite at 40°C gave 79 percent germination after six years of storage (Woodstock and Faul, 1981). Freeze-drying of seeds improves storability at higher temperatures. Seeds with 2.0 to 3.7 percent (ultradry) moisture preserve higher viability than those with 5.5 to 6.8 percent (dry) at 20°C storage, while no difference was observed at –20°C storage. It appears that ultradrying of seeds is useful, especially for seed storage under ambient conditions (Ellis et al., 1994, 1996).

Seed moisture. High seed moisture is injurious to the storage life of seeds. The rates of seed deterioration and fungal damage are greater at higher moisture contents. Seed moisture is higher at the initial stages of seed development, and it decreases on maturation and ripening. Seed moisture content depends on seed composition, air temperature, and relative humidity. At any given moisture level, seed viability and vigor decrease more rapidly at higher temperatures. Seedling vigor exhibits a rapid decline earlier than seedling viability (Doijode, 1990c). Harrington (1960b) reported that the seed longevity period doubled for every 1 percent decrease in moisture content. Low-moisture seeds are viable for longer periods (Minkov et al., 1974). According to Rocha (1959), loss of seed viability is rapid between 13 and 15 percent moisture content and slower at 11 percent. Low moisture decreases the seed deterioration, as it is unsuitable for the growth of pathogens (Christensen and Kaufman, 1965).

Seed Germination

Seed germination is rapid and high in onion at 20°C, whereas in another species, *Allium unifolium,* germination was high at 5°C, and seeds did not germinate at 15°C (Zimmer, 1994). Fresh seeds show a certain amount of dormancy. According to Lovato and Amaducci (1965), after-ripening of two months is required to eliminate the dormancy. Thus, dormancy tends to delay germination rather than completely prevent it. The cotyledon tip elongates during germination, carrying the radicle and plumule out of the testa. The elongated cotyledon forms a sharp bend that pushes to the soil surface, becomes green and straight, and produces foliage. Usik (1980) reported that large seeds give better field emergence than small seeds, and their yield also is higher by 18 percent. Higher concentrations of carbon dioxide inhibit germination (Thornton, 1944).

Storage Conditions

Storage conditions such as temperature, relative humidity, oxygen, light, and pathogens contribute to seed deterioration.

Temperature. Seed longevity decreases with an increase in storage temperatures. According to Harrington (1972), seed longevity decreased by half for every 5°C rise in storage temperature. The temperature is closely related to seed moisture, and the combination of these two factors seriously affects seed deterioration. High-moisture seeds at higher temperatures deteriorate rapidly. Szabo and Viranyl (1971) reported that onion seeds lose viability faster under variable temperatures and relative humidity. Kurdina (1966) noted a greater decrease in seed viability at 20°C and 80 percent RH. High-vigor seeds tolerate such effects (Barton, 1941). Different workers have reported various longevity periods at room temperatures. Thomazelli, Silva, and Sediyama (1990) observed that onion seeds are viable for 60 days or less under ambient conditions, and Bosewell et al. (1938) also noted a shorter storage life for onion seeds, which remained viable for less than 60 days at 26.7°C and 75 to 80 percent RH. According to Amaral and Bicca (1982), onion seeds remained viable for 12 months in an aluminum can, whereas Bacchi (1960) reported that seeds remained viable for four years at room temperatures when relative humidity was less than 10 percent. Seeds having 6.2 percent moisture remained viable for 13 years at room temperatures (Brown, 1939). Further, the reduction of seed moisture to 1.2 percent is beneficial for longer storage life (nine years) at room temperatures (Anonymous, 1954). In another study, seeds having 6 to 8 percent moisture, packed in polypropylene film or triple-laminated aluminum foil, and stored at 22°C and 40 to 60 percent RH remained viable for 65 months (Horky, 1991).

Seed deterioration is rather slow at lower storage temperatures, which are also noncongenial for insect and pathogen activities. Seed viability was lost more rapidly under higher storage temperature in high-moisture seeds than in low-moisture ones. In another study, seeds with 8 and 15 percent mc were packed in muslin cloth and in aluminum foil and stored at 5 and 30°C. Seeds in muslin cloth experienced moisture loss from 15 to 7 percent under high temperature and preserved high seed viability, but seed germination was zero within one week of storage in high-moisture seeds packed in aluminum foil pouches stored at 30°C (Dourado and Carson, 1994). Therefore, high-moisture seeds should be stored either in moisture-pervious containers or in unsealed conditions, especially under ambient storage conditions. Horky (1991) maintained onion seed viability for seven years by storing them at 0°C.

Subzero (–20°C) temperatures are most suitable for longer storage (Weibull, 1955; Zhang and Kong, 1996). Doijode (1998) preserved onion seeds for 15 years at 5 and –20°C. A higher percentage of germination was recorded in seeds stored at –20°C than in those stored at –5°C, particularly when seeds were stored in laminated aluminum foil pouches. Seedlings that emerged from stored seeds did not show any abnormalities, and they were on par with seedlings from fresh seeds for seed quality and morphological traits. Seeds stored for ten years at 5°C showed reduced germination percentage from 94 to 68 percent. Stanwood and Sowa (1995) noted greater seed deterioration at 5°C than at –18 or –196°C. Seeds stored at higher temperatures (40 to 50°C) and relative humidity (75 percent) showed certain unacceptable changes in seedlings, such as higher frequency of chromosomal aberrations. The disturbance of genetic stability was associated with loss of seed viability (Orlova, 1967).

Relative humidity. Humidity around the seeds regulates the seed moisture. Dry seeds gain moisture rapidly in high humidity conditions. Further, a high level of humidity favors the growth of storage fungi. When Bosewell and colleagues (1940) stored seeds at 44, 66, and 78 percent RH, percentage of germination was 73, 37, and 0 after 250 days of storage. Likewise, Barton (1943) reported that onion seeds lost viability rapidly at 55 and 70 percent RH, and they preserved high viability at 35 percent RH. Humidity fluctuation from 55 to 76 percent is more injurious to seed viability. To prevent the ill effects of high humidity, seeds were stored at a low humidity level and suitably packed in moisture-proof containers. Seeds stored at 10 percent RH gave 46 percent germination after five years of storage at 20°C, and they lost viability rapidly at RH of 40 percent and above (see Figure 36.3) (Doijode, unpublished data). Rusev and Bacvarov (1970) suggested storing dry seeds in sealed containers at room temperature with the enclosed air saturated with acetylene.

Air. The removal of air from storage containers enhances seed longevity. Seeds stored in partial vacuum gave 54 percent germination, as compared to those stored in air for 18 months under ambient conditions, which showed 27 percent germination (Doijode, 1988b), while no difference was noted when seeds were stored in air or partial vacuum at 34 to 40°C for 62 months (Brison, 1941). According to Lougheed and colleagues (1976), the deleteri-

FIGURE 36.3. Effect of Relative Humidity on Seed Longevity in Onion

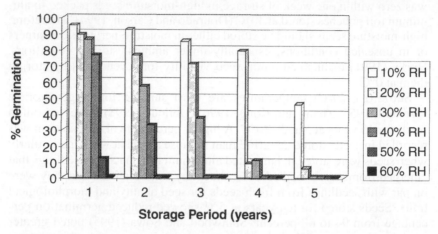

ous effects of high relative humidity decrease when onion seeds are stored under reduced atmospheric pressure for 14 weeks, and the moisture absorption is reduced by packing seeds in partial vacuum (Miyagi, 1966).

Light. Light hastens the germination process in certain vegetable crops, but darkness increases the germination percentage in most of the vegetable crops (Loy and Evensen, 1979). This is attributed to the role of phytochrome (Takai, Kendrick, and Dietrich, 1981). A 2 h exposure of seeds to fluorescent light every day prevents germination. The same seeds could be made germinable by subjecting them to dark conditions (Doijode, 1985a). Thanos and Mitrakos (1979) believed that intermittent light prevents the germination process more than continuous light, and inhibition is greater at higher light intensity (Ambica and Jayachandra, 1980). This is beneficial, especially in preserving seeds for a shorter period, because it inhibits sprouting during unfavorable storage conditions.

Storage fungi. Seeds are carriers of several pathogens on their surface and inside their tissue. The growth of these organisms is enhanced by the presence of high seed moisture and high relative humidity. Kurdina (1967) observed a greater number of mold fungi colonies at relative humidity above 70 percent and at 20°C storage. These pathogens cause seed discoloration, produce much of the heat in seeds, and reduce the germination percentage (Siddiqui, Majumdar, and Gaur, 1974). The fungi associated with onion seeds are *Alternaria alternata, Aspergillus flavus, A. niger, A. ochraceous, Chaetomium globosum, Curvularia lunata, Drechslera australiensis, Fusarium* spp., *Penicillium cyclopium, Rhizopus stolonifer,* and *Stachybotrys atra* (Gupta et al., 1984). The production of fungal metabolites reduces seed viability and vigor during storage (Gupta, Mehra, and Pandey, 1989). Use of fungicides effectively controls storage fungi. Gupta, Chhipa, and Bhargava (1982) reported that onion seeds remained viable (49 percent) up to a period of 40 months when treated with Ceresan and thiram (0.2 and 0.3 percent) followed by mancozeb (0.2 percent). Maintenance of low seed moisture (8 percent), low temperature (10°C), and RH less than 70 percent is beneficial for longer retention of pathogen-free seeds (Siddiqui, 1976).

Seed Storage Methods

Different storage methods, such as the use of desiccants, low temperature, and modified-atmosphere storage, maintain high viability and vigor for various periods. The methods should consider the principle factors associated with seed deterioration, such as temperature, humidity, oxygen, and pathogens, during storage. The choice of methods depends on the period of seed storage, kind and quantity of seed material to be stored, and cost. Maintaining genetic stability is the most important factor, apart from the retention of high viability and vigor in seeds. Seed storage begins with selection of bold, well-matured, healthy, and vigorous seeds for storage.

Storage with desiccants. In this method, seeds are stored in moisture-proof containers, such as glass bottles, plastic boxes, or aluminum cans or pouches, along with a known quantity of desiccant, such as silica gel. The desiccant absorbs excess moisture and maintains a low moisture level throughout the storage period. On moisture absorption, blue silica gel turns white. Drying in the sun or in an oven further dehydrates it. Silica gel has added advantages; it is easy to use, less expensive, and more effective in maintaining low seed moisture. According to Kruse (1997), silica gel should be added at the rate of 20 percent of seed weight for each 1 percent reduction in moisture content, while Grubben (1978) recommended application at the ratio of 1:10 (silica gel to seed quantity) in an airtight jar. Seeds preserved with silica gel retained higher viability for seven years when stored at both 5 and –20°C (see Figure 36.4) (Doijode, 1995). Silica gel is more useful at 5°C than at –20°C. Onion seeds stored in desiccators under warm conditions preserved viability for nine years in spite of fluctuations in temperature (Anonymous, 1954). Likewise, Harrison (1966) opined that addition of silica gel to the storage container maintains seed viability under adverse storage conditions, while Horky (1991) believed that the presence of a desiccant did not affect seed quality and that germination rate remain unchanged. However, silica gel does improve the seed quality in the presence of fluctuating storage temperatures. Calcium oxide (Barton, 1939) and dry rice are other desiccants used in onion storage (Currah and Msika, 1994). Dry rice preserves the onion seeds for three years at ambient temperatures.

FIGURE 36.4. Seed Germination As Influenced by Silica Gel During Storage of Onion Seeds

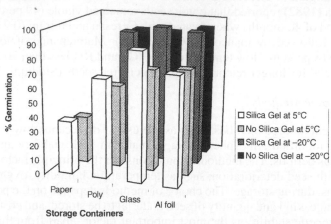

Seed packaging and sealed storage. Packaging is essential to protect the seeds from high humidity by providing a suitable moisture barrier structure. Many packaging materials, such as kraft paper bags, muslin cloth bags, glass containers, polyethylene pouches, aluminum cans, and laminated aluminum foil pouches, are used for seed storage. The choice depends on the kind and quantity of seed material, temperature, and period of storage. Cloth bags and kraft paper bags are suitable for short periods of seed storage. For longer storage, moisture-proof containers and low temperatures are employed. Harrington (1960a) reported that aluminum foil or tin cans are completely resistant to moisture penetration. Here, fluctuation of moisture occurs due to frequent opening and closing of the containers (Barton, 1953). Seeds of "Yellow Bermuda" onion remained viable for a longer period in sealed containers than in unsealed containers (Brison, 1953; Litynski 1966). In sealed storage, low-moisture seeds preserve their viability (Harrison and Carpenter, 1977), while high-moisture seeds (13 to 15 percent) could not maintain viability beyond 104 days when stored at 25°C (Rocha, 1959). Onion seeds stored in sealed jars at 15 to 20°C maintained storability longer than those stored in unsealed containers (Tronickova, 1965a). Aluminum foil pouches are completely impermeable to moisture, and seeds stored in them retain initial viability for 22 months (Miyagi, 1966). It is recommended that onion seeds be stored in aluminum foil pouches at 5°C (Boros and Hadnagy, 1968). These pouches are handy, occupy less space, and can be shaped to any size; they can be heat sealed and reused, and they do not release toxic compounds during storage. Elemery (1991) preserved onion seeds at 32 percent RH, and viability was unaffected by packaging or temperatures, whereas germination was significantly affected at 75 percent RH when seeds were stored in jute-woven poly-ethylene or cloth bags, but not in polyethylene film covered with jute or cloth bags. Seed moisture content increases if seeds are stored in jute or cloth bags. Onion seeds having 6.5 percent moisture content and packed in paper packets and in laminated aluminum foil pouches were stored under ambient conditions (16 to 35°C and 25 to 90 percent RH). Seed viability is well preserved in sealed storage and decreases rapidly under unsealed conditions (see Figure 36.5) (Doijode, 1987). Laminated aluminum foil pouches and polyethylene bags are effective in preserving high seed viability for five years at 5 and −18°C (Doijode, 1989). Viability was further extended to a period of 15 years by storage at −20°C (Doijode, 1998). A well-dried onion seed can be stored for a longer period in sealed storage.

Low-temperature storage. Low-temperature storage is the most ideal method for onion seed storage. Seeds remain viable for fairly longer periods at low temperatures. The storage temperature is selected based on purpose and length of seed storage. High-quality seeds are preserved at subzero or ultralow temperatures for longer periods, while for bulk storage, seeds are stored above 0°C with a dehumidifier for shorter periods. Seeds are to be well dried for subzero temperature storage; otherwise, they will be affected by chilling injury. Moisture-proof containers such as metal cans, glass containers, and aluminum foil pouches are used for these purposes.

Seed deterioration is significantly reduced at lower temperatures. Cold storage of onion seeds is a very satisfactory method for conservation of high seed viability for longer periods (Doijode, 1987). Onion seeds were successfully preserved at −20°C for 15 years without decline in viability and vigor,

FIGURE 36.5. Seed Longevity During Open and Sealed Storage of Onion Seeds Under Ambient Conditions

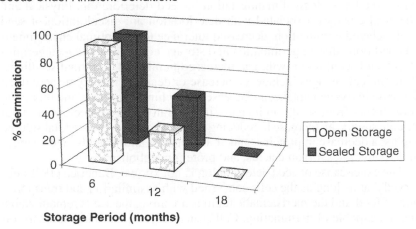

% Germination

Storage Period (months)

□ Open Storage
■ Sealed Storage

and there were no morphological variations in the seedlings that emerged from cold-stored seeds (Doijode, 1998). According to Tronickova (1965b), seed germination and seedling emergence are improved following sealed storage at −14 to −20°C. Barton (1953) stored onion seeds at −4°C for 20 years. Likewise, Gane (1948) observed high viability at 0 and 10°C up to 60 percent RH, and −20°C is superior to 0 or 5°C (Weibull, 1952, 1955). Rusev (1974) stored onion seeds in sealed glass containers at −1 to −2°C for three to four years.

Modified-atmosphere storage. Seed viability is preserved well in the absence of oxygen. Replacement of oxygen with other inert gases, such as carbon dioxide or nitrogen, is beneficial in maintaining high viability, especially at lower seed moisture levels. A high quantity of oxygen during storage affects seed viability. According to Roberts and Abdalla (1968), the deleterious effects of oxygen are more pronounced at high levels of seed moisture. Harrington (1960a) reported that onion seeds deteriorated faster when stored in oxygen, and slower when stored in nitrogen. Seeds stored in a nitrogen atmosphere showed high seed viability (Doijode, unpublished data), and longevity improves with seed storage in nitrogen, more so than in carbon dioxide (Doijode, unpublished data). However, onion seed storage in carbon dioxide is more beneficial than storage in air (Lewis, 1953; Harrison and Mcleish, 1954). Harrison (1956) reported that carbon dioxide-stored seeds preserved high viability for six years. Gaseous storage is effective for short-term storage and beneficial in the absence of a cold-storage facility.

Invigoration of Stored Seeds

Seed deterioration. Onion seeds are highly perishable (Ellis and Roberts, 1981), and reductions in germination and seedling vigor are the major criteria for the physiological manifestation of seed deterioration. Physical and chemical changes associated with deterioration are discoloration of seed coat, delayed germination, decreased tolerance to suboptimal environmental conditions during germination and storage conditions, reduced germinability and seedling growth, and increased number of abnormal seedlings. Biochemical changes include an increase or decrease in enzyme activity, a decrease in oxygen uptake, an increase in leaching of organic and inorganic constituents from seeds, an increase in free fatty acids, a decrease in total soluble sugars, an increase in reducing sugars and a decrease in total soluble sugars, a decrease in proteins, and an increase in amino acids, and changes in carbohydrates, organic acids, and protein metabolism.

The exact cause of seed deterioration is not known. Banerjee (1978) observed that as long as the cells concerned with assimilation and transportation of food and the meristematic regions are alive, the seeds remain viable and are capable of germinating. Cell structure gets disturbed on deterioration, resulting in excessive leakage of electrolytes, soluble sugars, and potassium (Doijode, 1985b). In onion seeds, with increasing storage period, the palmitic and stearic acids increase and the oleic and linoleic acids decrease. According to Suh, Lee, and Park (1996), seed viability is negatively correlated with saturated fatty acid content and positively correlated with unsaturated fatty acid content. Seed vigor, an essential component for crop production and better storage, decreases on deterioration (Perry, 1981).

Normally, high seed vigor is estimated by percentage of germination. Kononkov, Kononkova, and Startsev (1991) developed a rapid technique for estimating seed quality. Seeds are kept in gauge bags placed in boiling water for 30 min. The number of seeds producing radicles is directly proportional to seed viability. Based on estimated storage conditions, Ellis and Roberts (1980) developed seed viability equations to predict seed storability. Stumpf, Peske, and Baudet (1997) showed that such equations are less accurate at higher levels of moisture and for longer storage periods. The possibility of using ATP (adenosine triphosphate) content as an indicator of onion seed quality was also studied (Siegenthaler and Douet-Orhant, 1994). A high positive correlation exists between ATP content and viability. ATP content is greater for seeds stored at 3°C than for those stored at 30°C. Likewise, dehydrogenase activity, which is directly related to seed viability, is greater in seeds stored at low temperatures than in those stored under ambient conditions (Doijode, 1990a). Exposing seeds to accelerated aging (42 or 45°C at 100 percent RH) for 72 h can help predict seed quality (Amaral, 1983). Seed quality can be improved to a certain extent by imposing the following treatments.

Seed priming. Seeds have to be treated with osmotic solution to hasten the germination process. This gives rapid and uniform seedlings and delays aging (Dearman, Brocklehurst, and Drew, 1986). PEG is most effective in this regard (Brocklehurst, Dearman, and Drew, 1987). Primed seeds are dried to their original level of moisture and stored. This gives higher germination than unprimed seeds, a result attributed to the cellular repair process during the imbibition state (Burgass and Powell, 1984). According to Basu (1990), prestorage dry treatments, such as acetone permeation of antiaging chemicals and low concentrations of halogens (iodine, chlorine, and bromine) and alcohol (methanol, ethanol, and isopropanol), improve the storability of dry-stored seeds.

The intricacies involved in the invigoration of seeds are not clear. It appears that prevention of damaging oxidative reactions, such as free-radical-induced peroxidation reactions involving unsaturated lipid molecules of lipoprotein biomembranes, and the repair of age-induced damage to vital bioorganelles by the cellular repair system are associated with seed invigoration.

Midstorage treatment with water or saline solution helps in retaining the original vigor and viability of seeds (Basu, 1976). Doijode and Raturi (1987) reported that hydration and dehydration improve the storage life and vigor of seeds. This improvement is associated with enhanced activity of dehydrogenase and peroxidase, with a simultaneous reduction in lipolytic activity and breakdown of fat contents. Onion seeds were soaked in water and other solvents, such as ethyl alcohol, acetone, xylene, dichloromethane, and petroleum ether. Seeds soaked in water followed by drying gave higher germination than other solvents (Banerjee, Negi, and Jain, 1982). High-moisture onion seeds (15 to 25 percent) were stored at high temperatures (45°C); the loss of viability was less for seeds having moisture of 20 to 25 percent than for those having 15 percent moisture. It is suggested that activation of the cellular repair process is due to imbibitions (Ward and Powell, 1983). Addition of monosodium hydrogen phosphate (10^{-4} M) and sodium sulfate (10^{-4} M) to water used for soaking gives minor additional advantages (Choudhuri and Basu, 1988).

Chemicals. Chemical treatment of seeds is beneficial in delaying the aging process. Impregnation of acetone with vitamin E (20 units/milliliter [ml]) and butylated hydroxytolule (10^{-1} M) improves the storability of onion seeds. Natural and synthetic antioxidants can also be used, either alone or in combination with drying, to extend seed longevity (Woodstock et al., 1983). Six-month-old onion seeds treated with certain chemicals, such as EDTA (ethylenediaminetetraocetic acid) (0.1 M), potassium metabisulfite (KMS) (0.02 M), oxalic acid (0.01 M), ascorbic acid (0.02 M), PEG (0.5 M), and saturated glucose can be stored under ambient conditions (16 to 35°C and 25 to 90 percent RH). Seed germination and vigor improved with the treatment. These seeds were viable for a longer period than untreated seeds during ambient storage (see Figure 36.6) (Doijode, 1988a). These chemicals act as antioxidants or prevent free-radical formation. Leaching of toxic metabolites from seeds occurs during imbibitions. According to Dasgupta, Basu, and Basu (1976), leaching of inhibitors is greater in the presence of dilute salt. EDTA prevents the formation of radical centers on unsaturated fatty acids and prevents the deterioration of seeds (Demopoulous, 1973). EDTA also extends seed storage life and vigor through the cellular repair system during the hydration phase (Villiers and Edgcumbe, 1975).

Growth regulators. Seed quality improves with the application of growth regulators. Kinetin improved germination percentage and seedling vigor even in seeds stored under unfavorable storage conditions, while ABA lowered seed viability and vigor (Styer and Cantliffe, 1977).

Magnetic stimulus. Electromagnetic treatment of seed causes morphological, physiological, and biochemical changes in seeds (Pittman and Ormrod, 1970). Five-year-old onion seeds stored at 5°C can be exposed to

FIGURE 36.6. Influence of Certain Chemicals on Onion Seed Viability After Nine Months of Ambient Storage

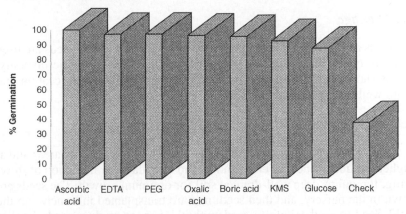

electromagnetic fields (36 to 144 oersteds). Seed germination is improved with exposure to a 108 oersteds electromagnetic field for 30 min. This also improved seedling length and dry weight, and leaching of metabolites decreased as compared to untreated control (Alexander and Doijode, 1995). Magnetic stimulus promotes the biological mechanism by slowing down the process of seed deterioration and stimulates the seedling growth rate.

LEEK: *Allium ameloprasum* var. *porrum* L.

Introduction

Leek is a robust plant popular in the Middle East and Northern Europe. It does not form bulbs; instead, the long green leaves are cooked as a vegetable. The blanched stems and leaves are boiled or fried. It is a rich source of minerals such as phosphorus and iron.

Origin and Distribution

Leek originated in the Mediterranean region. It is cultivated in Belgium, China, Denmark, France, Malaysia, the Netherlands, the Philippines, Spain, and the United Kingdom.

Morphology

Leek belongs to family Alliaceae and has chromosome number of 2n = 32. It is a monocotyledonous biennial plant. Leaves are flattened and variable in length. Inflorescence is a terminal umbel with many pinkish flowers. Fruit is a capsule. One thousand seeds weigh about 3.75 g.

Seed Storage

Seeds are generally used in the propagation of leek. Also, seeds are used in crop improvement and in long-term conservation of genetic diversity. Seeds show orthodox storage behavior; they can tolerate desiccation and store well at chilling temperatures.

Seed Collections

Leek is a cool-season crop. It grows well in low-temperature areas and at higher altitudes. It needs vernalization and long days for induction of flowering. It requires deep well-drained soil for optimum growth. The seeds are sown in the nursery, and then seedlings are transplanted in trenches in the field. Seed-to-seed or bulb-to-seed method is used to produce seeds. Leek is a cross-pollinated crop, and bees bring about pollination. An isolation distance of 1,000 m is maintained between two different varieties. Normal cultural practices, such as fertilizer, irrigation, and plant protection, are followed. Off-type plants are removed. Mature heads are collected, dried, and threshed; seeds are separated, cleaned, and packed for storage.

Seed Germination

Seed germination is rapid at low temperature. It is highest at 20°C and is affected at 30°C (Gelmond, 1965). Likewise, Corbineau, Picard, and Come (1994) reported that the optimum temperature for germination was between 15 and 20°C. Fresh seeds exhibit dormancy that persists for several months. Prechilling of seeds and exposure to an alternate temperature regime (20/30°C for 16/8 h) promotes germination.

Storage Conditions

Leek seeds are short-lived under ambient conditions. Seed viability and vigor decrease with an increase in storage temperature and relative humidity. Seeds stored at 26.7°C gave 65 percent germination after six years of storage, and storage at 43.3°C further reduced seed viability (Adamson, 1960). Litynski and Chudoba (1969) preserved high viability for two years at 2°C and 10 percent RH. Further, seeds with 7.7 to 13.8 percent mc stored at −20°C retain their original viability for five years (Tronickova, 1965a). The stimulatory effect of priming persists after drying and subsequent storage with silica gel up to 15 months (Corbineau, Picard, and Come, 1994).

REFERENCES

Adamson, R.M. 1960. The effect on germination of drying leek seeds heads at different temperatures. *Can. J. Plant Sci.* 40:666-671.

Agrawal, P.K. 1976. Identification of suitable storage places in India on the basis of temperature and relative humidity conditions. *Seed Res.* 4:6-11.

Agrawal, P.K. 1980. Relative storability of seeds of ten species under ambient conditions. *Seed Res.* 8:94-99.

Alexander, M.P. and Doijode, S.D. 1995. Electromagnetic field a novel tool to increase germination and seedling vigor of conserved onion *(Allium cepa)* and rice *(Oryza sativa)* seeds with low viability. *Plant Genetic Resources Newsletter* 104:1-5.

Amaral, A.Do.S. 1983. Early aging a test of seed vigour. *Lavoura Arrozeira* 36: 24-25.

Amaral, A.Do.S. and Bicca, L.H.F. 1982. Storage of onion seeds in sealed cans. *Lavoura Arrozeira* 35:16-20 .

Ambica, S.R. and Jayachandra. 1980. Influence of light on seed germination in *Eupatorium odoratum. Indian Forester* 106:637-640.

Anonymous. 1954. The preservation of viability and vigor in vegetable seeds. *Asgrow Monograph New Haven* No. 2:32.

Bacchi, O. 1960. Seed storage studies. VI. Onion. *Bragantia* 19:99-102.

Banerjee, S.K. 1978. Observations on the initiation of seed deterioration and its localisation in barley and onion. *Seed Sci. Technol.* 6:1025-1028.

Banerjee, S.K., Negi, H.C.S., and Jain, S.K. 1982. Solvents affecting seed germination and seedling growth. *Seed Res.* 10:204-208.

Barton, L.V. 1935. Storage of vegetable seeds. *Boyce Thompson Inst. Contrib.* No. 7:323-332.

Barton, L.V. 1939. A further report on the storage of vegetable seeds. *Boyce Thompson Inst. Contrib.* No. 10:205-220.

Barton, L.V. 1941. Relation of certain air temperatures and humidity to viability of seeds. *Boyce Thompson Inst. Contrib.* No. 12:85-102.

Barton, L.V. 1943. Effect of moisture fluctuations on the viability of seeds in storage. *Boyce Thompson Inst. Contrib.* No. 13:35-45.

Barton, L.V. 1953. Seed storage and viability. *Boyce Thompson Inst. Contrib.* No. 17:87-103.

Basu, R.N. 1976. Physico-chemical control of seed deterioration. *Seed Res.* 4:15-23.

Basu, R.N. 1990. Seed invigoration for extended storability. *Inter. Conf. Seed Sci. Technol. New Delhi.* Abstr. 2.11:38.

Behairy, A.G. and El-Habbasha, K.M. 1979. Onion seed production as affected by vernalization of bulbs. *Zeitschriftf Acker und Pflanzenbau* 148:109-114.

Boros, D. and Hadnagy, A. 1968. Storage of onion and cabbage seeds. *Zoldsegtermesztes* 2:81-90.

Bosewell, V.R., Toole, E.H., Toole, V.K., and Fisher, D.F. 1938. A study of rapid deterioration of vegetable seeds and methods for its prevention. *Proc. Am. Soc. Hort. Sci.* 36:655-659.

Bosewell, V.R., Toole, E.H., Toole, V.K., and Fisher, D.F. 1940. A study of rapid deterioration of vegetable seeds and methods for its prevention. *U.S.D.A. Tech. Bull.* 708:1-40.

Brison, F.R. 1941. Influence of storage conditions upon the germination of onion seeds. *Tex. Acad. Sci. Proc. Trans.* 25:65-71.

Brison, F.R. 1953. Influence of storage conditions on the germination of onion seeds. *Prog. Rep. Texas Agric. Exp. Stat.,* p. 4.

Brocklehurst, P.A., Dearman, J., and Drew, R.L.K. 1987. Recent developments in osmotic treatment of vegetable seeds. *Acta Hort.* 215:193-200.

Brown, E. 1939. Preserving the viability of Bermuda onion seeds. *Science* 89:292-293.

Burgass, R.W. and Powell, A.A. 1984. Evidence for repair process in the invigoration of seeds by hydration. *Ann. Bot.* 53:753-757.

Choudhuri, N. and Basu, R.N. 1988. Maintenance of seed vigor and viability of onion *(Allium cepa). Seed Sci. Technol.* 16:51-61.

Christensen, C.M. and Kaufman, H.H. 1965. Deterioration of stored grains by fungi. *Ann. Rev. Phytopathol.* 3:69-84.

Copeland, L.O. 1976. *Principles of seed science and technology.* Minneapolis, MN: Burgess Publishing Co.

Corbineau, F., Picard, M.A., and Come, D. 1994. Germinability of leek seeds and its improvement by osmopriming. *Acta Hort.* 371:45-52.

Currah, L. and Msika, L. 1994. Cheap methods of storing small quantities of onion seeds without refrigeration. *Onion Newsletter Tropic.* No. 6:61-62.

Dasgupta, M.P., Basu, P., and Basu, R.N. 1976. Seed treatment for vigour, viability and productivity of wheat *(Triticum aestivum* L.). *Indian Agric.* 20:265-273.

Dearman, J., Brocklehurst, P.A., and Drew, R.L.K. 1986. Effect of osmotic priming and aging on onion seed germination. *Ann. Appl. Biol.* 108:639-648.

Demopoulous, H.B. 1973. Control of free radicals in biologic system. *Fed. Proc.* 32:1903-1908.

Doijode, S.D. 1985a. Influence of irradiance on seed germination. *Die Gartenbauwissenschaft* 50:29-30.

Doijode, S.D. 1985b. Onion seed quality in relation to seed deterioration. *Veg. Sci.* 12:59-63.

Doijode, S.D. 1987. Effect of packaging material and storage temperatures on the longevity of onion seeds. *Proc. Natl. Symp. Pl. Genetic Resources* 1:57.

Doijode, S.D. 1988a. Effect of pre-treatment on the germination of onion *(Allium cepa)* seeds. *Die Gartenbauwissenschaft* 53:101-102.

Doijode, S.D. 1988b. Studies on partial vacuum storage of onion *(Allium cepa)* and bell pepper *(Capsicum annuum* L.) seeds. *Veg. Sci.* 15:126-129.

Doijode, S.D. 1989. Effect of storage temperature on longevity of onion seeds. *Indian J. Pl. Genetic Resources* 2:41-43.

Doijode, S.D. 1990a. Conservation of onion *(Allium cepa)* germplasm. *Proc. Natl. Sem. Onion Garlic Solan* 1:44-46.

Doijode, S.D. 1990b. Germinability and longevity of onion seeds under different drying methods. *Prog. Hort.* 22:117-120.

Doijode, S.D. 1990c. Influence of storage conditions on germination of onion seeds. *J. Mah. Agric. Universities* 15:34-35.

Doijode, S.D. 1990d. Solute leakage in relation to loss of seed viability in onion cultivars. *Indian J. Pl. Physiol.* 33:54-57.

Doijode, S.D. 1995. Effect of silica gel and storage containers on seed viability and vigor in onion *(Allium cepa). Seed Res.* 23:67-70.

Doijode, S.D. 1998. Conservation of *Allium cepa* germplasm in India. *Natl. Symp. Veg. Varanasi.* Abstr. No. 3:2.

Doijode, S.D. and Raturi, G.B. 1987. Effect of hydration and dehydration on viability and vigor of certain vegetable seeds. *Seed Res.* 15:156-159.

Dourado, A.M. and Carson, R. 1994. Experimental storage of onion *(Allium cepa)* seed under simulated tropical conditions. *Onion Newsletter Tropic* No. 6:53-55.

Duvel, J.W.T. 1905. The vitality of buried seeds. *U.S.D.A. Bur. Pl. Indus. Bull.* 83:20.

Elemery, M.I. 1991. Effect of storage conditions (Temp°C and RH percent) and packing materials on germination percent of onion and sunflower seeds. *Ann. Agril. Sci.* 29:657-667.

Ellis, R.H., Hong, T.D., Astley, D., and Kraak, H.L. 1994. Medium term storage of dry and ultra dry seeds of onion at ambient and sub-zero temperatures. *Onion Newsletter Tropic* No. 6:56-58.

Ellis, R.H., Hong, T.D., Astley, D., Pinnegar, A.E., and Kraak, H.L. 1996. Survival of dry and ultra dry seeds of carrot, groundnut, lettuce, oilseed rape and onion during five years of hermetic storage at two temperatures. *Seed Sci. Technol.* 24:347-358.

Ellis, R.H. and Roberts, E.H. 1977. A revised seed viability nomograph for onion. *Seed Res.* 5: 93-95.

Ellis, R.H. and Roberts, E.H. 1980. Improved equations for the prediction of seed longevity. *Ann. Bot.* 50:69-82.

Ellis, R.H. and Roberts, E.H. 1981. The quantification of aging and survival in orthodox seeds. *Seed Sci. Technol.* 9:373-409.

Gamiely, S., Smitte, D.A., and Mills, H.A. 1990. Onion seed size, weight and elemental content affect germination and bulb yield. *Hort. Sci.* 25:522-523.

Gane, R. 1948. The effect of temperature, water content, and composition of the atmosphere on the viability of carrot, onion, and parsnip seeds in storage. *J. Agril. Sci.* 38:84-89.

Gelmond, H. 1965. Pretreatment of leek seeds as a means of overcoming super optimal temperatures of germination. *Proc. Int. Seed Test. Assoc.* 30:737-742.

Gray, D. and Steckel, J.R.A. 1984. Viability of onion *(Allium cepa)* seeds as influenced by temperatures during seed growth. *Ann. Appl. Biol.* 104:375-382.

Grubben, G.J.H. 1978. Vegetable seeds for the tropics. *Royal Trop. Inst. Amsterdam Bull.* 301:40.

Gupta, R.B.L., Chhipa, H.P., and Bhargava, A.K. 1982. Role of seed dressing fungicides in determining longevity of some vegetable seeds. *Indian J. Mycol. Pl. Path.* 12:116.

Gupta, R.P., Mehra, U., and Pandey, U.B. 1989. Effect of various chemicals on viability of onion seeds in storage. *Seed Res.* 17:99-101.

Gupta, R.P., Srivastava, P.K., Srivastava, V.K., and Pandey, U.B. 1984. Note on the fungi associated with onion seeds, their pathogenicity, and control. *Seed Res.* 12:98-100.

Harrington, J.F. 1960a. Preliminary report on the relative desirability of different containers for storage of several kinds of vegetable seeds. *Calif. Coll. Agr. Veg. Crops* Ser. No. 104:8.

Harrington, J.F. 1960b. Thumb rule of drying seeds. *Crops Soils* 13:16-17.

Harrington, J.F. 1972. Seed storage and longevity. In *Seed biology,* Volume III, Kozolowski, T.T. (Ed.). New York: Academic Press, pp. 145-245.

Harrison, B.J. 1956. Seed storage. *John Innes. Hort. Inst. Ann. Rep.* 46:15-16.

Harrison, B.J. 1966. Seed deterioration in relation to storage conditions and its influence upon germination, chromosomal damage and plant performance. *J. Natl. Inst. Agric. Bot.* 10:644-663.

Harrison, B.J. and Carpenter, R. 1977. Storage of *Allium cepa* seeds at low temperature. *Seed Sci. Technol.* 5:699-702.

Harrison, B.J. and Mcleish, J. 1954. Abnormalities of stored seeds. *Nature* 173:593-594.

Hesse, P.S., Vest, G., and Honma, S. 1979. The effect of four storage treatments on seed yield components of three onion inbreds. *Scientia Hort.* 11:207-215.

Horky, J. 1991. The effect of temperatures on the long term storage of dry seeds of some selected vegetables. *Zahradnictvi.* 18:29-33.

Katan, J., Abramski, M., and Levi, D. 1975. Transport and transmission of *Fusarium* pathogens by tomato and onion seeds. *Phytoparasitica* 3:74.

Kononkov, P.F., Kononkova, S.N., and Startsev, V.N. 1991. Determination of viability and predicting longevity of onion seeds. *Sel'skokhozyaistvennaya Biol.* No. 1:187-195.

Kruse, M. 1997. Store seeds with blue gel. *Genuise* 33:88-89.

Kurdina, V.N. 1966. Changes in the sowing quality of seeds of vegetable crops during storage. *Izv timirjazev sel hoz Akad* No. 5:135-144.

Kurdina, V.N. 1967. Changes in the microflora and viability of vegetable seeds during storage. *Izv timirjazev sel hoz Akad* No. 2:175-183.

Lewis, D. 1953. Seed storage. A.R. *John. Innes Hort. Inst.* 43:14-15.

Litynski, M. 1966. Long term onion seed storage without the use of low temperature. *Proc. Inter. Seed Test. Assoc.* 31:223-231.

Litynski, M. and Chudoba, Z. 1969. The effect of temperature on the germinating capacity of parsley and leek seeds under different humidity levels when stored in airtight containers. *Biul. Warzyw* 9:327-335.

Lougheed, E.C., Murr, D.P., Harney, P.M., and Sykes, J.T. 1976. Low pressure storage of seeds. *Experientia* 32:1159-1161.

Lovato, A. and Amaducci, M.T. 1965. Examination of the problem of whether dormancy exists in seeds of onion (*Allium cepa* L.) and leek (*Allium porrum* L.). II. Effect of temperature, prechilling and light on germination. *Proc. Inter. Seed Test. Assoc.* 30:803-820.

Loy, J.B. and Evensen, K.B. 1979. Phytochrome relation of seed germination in dwarf strain of watermelon. *J. Am. Soc. Hort. Sci.* 104:496-499.

Minkov, I., Ivanov, L., Rusev, D., and Gadzhonova, P. 1974. Investigation on storage conditions for certain vegetable seeds and the determination of their natural weight losses. *Gradinarska i Lozarska Nauka* 11:49-58.

Miyagi, K. 1966. Effect of moisture-proof packing on the maintenance of viability of vegetable seeds. *Proc. Inter. Seed Test. Assoc.* 31:213-222.

Orlova, N.N. 1967. Study of mutation processes in dormant seeds of welsh onion stored under condition of high temperature and humidity. *Genetika* 3:19-25.

Perry, D.A. 1981. Handbook of vigor test methods. *Inter. Seed Test. Assoc. Zurich.*

Pittman, U.J. and Ormrod, D.P. 1970. Physiological and chemical features of magnetically treated winter wheat seedling. *Can. J. Plant Sci.* 50:211-217.

Pohler, H. 1984. Seed drying with dehumidified air. *Wissenschaftliche Beitrage* 54:482-490.

Pritchard, E.W. 1933. How long do seeds retain their germinability power? *Dept. Agril. Aust. Rep. Agr. J.* 36:645-646.

Roberts, E.H. and Abdalla, F.H. 1968. The influence of temperature, moisture and oxygen on period of seed viability in barley, broad bean and peas. *Ann. Bot.* 32:97-117.

Rocha, F.F. 1959. Interaction of moisture content and temperature on onion seed viability. *Proc. Am. Soc. Hort. Sci.* 73:385-389.

Rusev, D. 1974. The effects of the conditions and length of storage on the viability of onion seeds. *Grad. Lozar. Nauka.* 11:41-48.

Rusev, D. and Bacvarov, S. 1970. The possibility of prolonged onion seed storage. *Grad. Lozar. Nauka.* 7:45-53.

Sandhu, J.S., Nandpuri, K.S., and Thakur, J.C. 1972. Studies on the freshly harvested mature and immature onion seeds. *Indian J. Hort.* 29:339-341.

Siddiqui, M.R. 1976. Pathology of stored seeds. *Seed Res.* 4:66-72.

Siddiqui, M.R., Majumdar, A., and Gaur, A. 1974. Fungal flora associated with the seeds of cereals and vegetables in India. *Seed Res.* 2:46-50.

Siegenthaler, P.A. and Douet-Orhant, V. 1994. Relationship between the ATP content measured at three imbibitions times and germination of onion seeds during storage at 3, 15 and 30°C. *J. Exp. Bot.* 45:1365-1371.

Stanwood, P.C. and Sowa, S. 1995. Evaluation of onion *(Allium cepa)* seeds after ten years of storage at 5, −18°C and −196°C. *Crop Sci.* 35:852-856.

Stumpf, C.L., Peske, S.T., and Baudet, L. 1997. Storage potential of onion seeds hermetically packaged at low-moisture content. *Seed Sci. Technol.* 25:25-33.

Styer, R.C. and Cantliffe, D.J. 1977. Effect of growth regulators on storage life of onion seeds. *Proc. Fl. St. Hort. Soc.* 90:415-418.

Suh, J.Y., Lee, D.J., and Park, K.W. 1996. Germination characteristics and fatty acid composition as affected by storage period in onion seeds. *RDA. J. Agril. Sci. Hort.* 38:448-453.

Szabo, L. and Viranyl, S. 1971. Investigation on the germination of cultivated plant seeds stored under variable conditions. *Agrobotanika* 12:15-20.

Takai, M., Kendrick, R.E., and Dietrich, S.M.C. 1981. Interaction of light and temperature on the germination of *Rumex obtusifolius. Planta* 152:209-214.

Thanos, C.A. and Mitrakos, K. 1979. Phytochrome mediated germination control of maize caryposes. *Planta* 146:415-417.

Thomazelli, L.F., Silva, R.F.D., and Sediyama, C.S. 1990. How to preserve onion seed quality. *Agropecuaria Caterinense* 3:7-8.

Thornton, N.C. 1944. Carbon dioxide storage. XII. Germination of seeds in the presence of CO_2. *Boyce Thompson Inst. Contrib.* 13:355-360.

Tronickova, E. 1965a. The effect of sub-zero temperatures on the biology of some vegetable seeds. *Ved. Pr. ustred. vyzk. Ust. rostl Vyroby v praze Ruzyni* 9:27-39.

Tronickova, E. 1965b. Storage of onion seeds at sub-zero temperature and in sealed containers. *Bull. Vyzk. Ust. Zelin. Olomauc* 9:121-129.

Usik, G.E. 1980. Effect of seed size on emergence and yield in onion and carrot. *Intensifik Ovosch Kishinev Mold,* pp. 30-33.

Villiers, T.A and Edgcumbe, D.J. 1975. On the cause of seed deterioration in dry storage. *Seed Sci. Technol.* 3:761-774.

Ward, F.H. and Powell, A.A. 1983. Evidence of repair process in onion seeds during storage at high seed moisture contents. *J. Exp. Bot.* 34:277-282.

Weibull, G. 1952. The cold storage of vegetable seeds and its significance for plant breeding and the seed trade. *Agri. Hortique Genetica.* 10:97-104.

Weibull, G. 1955. The cold storage of vegetable seeds—Further studies. *Hort. Genet.* 13:121-142.

Woodstock, L.W. 1975. Freeze drying as an alternate method for lowering seed moisture. *Proc. Assoc. Off. Seed Ana.* 65:159-163.

Woodstock, L.W. and Faul, K. 1981. Extended storage of freeze dried onion, sweet pepper and parsley seeds. *Newsletter Assoc. Off. Seed Ana.* 55:15.

Woodstock, L.W., Maxon, S., Faul, K., and Bass, L.N. 1983. Use of freeze drying and acetone impregnation with natural and synthetic antioxidants to improve storability of onion, pepper and parsley seeds. *J. Am. Soc. Hort. Sci.* 108:692-696.

Woodstock, L.W. and Schroeder, E.M. 1975. Longer life for stored seeds. *Agril. Res.* 23:8-9.

Woodstock, L.W., Simkin, J., and Schroeder, E. 1976. Freeze drying to improve seed storability. *Seed Sci. Technol.* 4:301-311.

Zhang, H.Y. and Kong, X.H., 1996. Tests on effect of vegetable seeds stored for many months. *Acta-Agri. Bore-Sinica* 11:118-123.

Zimmer, K. 1994. Germination of two American *Allium* species. *Die Gartenbauwissenschaft* 59:125-126.

37

Root Crops

RADISH: *Raphanus sativus* L.

Introduction

Radish is a popular vegetable crop in Asia and Europe. It is mainly cultivated for the young tuberous root, which is used in salad. Tender and nonpithy roots are cooked as a vegetable and also made into pickles. The foliage is also used as a leafy vegetable in Asia. The immature pods and leaves of another species, *Raphanus caudatus,* are eaten raw and/or cooked as a vegetable. Radish increases the appetite and gives a cooling effect. Roots are rich in minerals and vitamins A and C. The pungency of radish root is due to a volatile substance called isothiocyanate. Radish has medicinal properties, especially in curing ailments of the liver and gall bladder, headache, diarrhea, piles, and jaundice.

Origin and Distribution

Radish probably originated in Western Asia, including the greater portion of China, where wide genetic diversity exists. Later, its cultivation spread to temperate and tropical regions. It is grown in Brazil, Germany, Hungary, India, Indonesia, Italy, Japan, Malaysia, the Netherlands, the Philippines, Spain, the United Kingdom, and the United States.

Morphology

Radish belongs to family Cruciferae and has chromosome number 2n = 18. It is dicotyledonous annual or biennial herb. Its hypocotyls and swollen primary roots are used in salad. Root size, shape, and color vary in different cultivars. They are round, tapering, and cylindrical, either red or white, and flesh is white. Leaves are pinnate and partly divided. Flowers are small, white, or pink, self-incompatible, and pollinated by insects. Fruit is an indehiscent silique, with a long beak, and contains 6 to 12 seeds. Seeds are small, 3 mm in diameter, yellow or brown, and globose. One thousand seeds weigh about 10 g.

Seed Storage

Radish is raised commercially through seeds. Seeds show orthodox storage behavior and are capable of maintaining high viability, vigor, and genetic stability on long-term conservation. Radish seeds are fairly good storers under ambient conditions.

Seed Collections

Radish is a cool-season crop. Asian types are suited to tropical conditions; European ones are less pungent and grow well in tropical plains, and produce seeds in hilly regions. Small radishes are adapted to temperate climates, and larger ones can tolerate fluctuations in temperature. Low temperature is congenial for development of root texture, flavor, and size. High temperature leads to premature bolting. Radish prefers loose, fertile, and well-drained soils. Seeds are produced either by root-to-seed or seed-to-seed method. In root-to-seed method, the matured roots, called stecklings, are removed and replanted by cutting the tops. Seed size and weight decide seed quality. With increased seed size, the field emergence increases (Pandita and Randhawa, 1992). Large seeds also give higher yield than smaller seeds (Lee and Nichols, 1978). Proper cultural practices, such as adequate spacing, nutrition, irrigation, and plant protection measures, are to be followed for production of quality seeds. Radish is a cross-pollinated crop, and bees and other insects bring about pollination. Thus, a minimum isolation distance of 1,600 m is maintained between cultivars. The off-type roots are rogued based on shape and color. The crop is harvested when pods become brown and seeds near maturity. Dried pods are collected and threshed; seeds are removed, dried, cleaned, and packed for storage.

Seed Germination

Fresh seeds show dormancy, and six weeks of after-ripening eliminates the dormancy (Tokumasu, 1971). Further, germinating seeds using an alternate temperature regime of 10/20°C for 18/6 h breaks the dormancy (Mekenian and Millemsen, 1975). In radish, seed germination is epigeal and takes five days to complete. Fresh seeds absorb more water and germinate more slowly than older seeds (Vasil'eva and Lazukov, 1970). Seed germination improves by application of gibberellic acid (5 to 10 ppm) (Pawar, Joshi, and Mahakal, 1977), naphthaleneacetic acid (10 to 20 ppm) (Singh and Dohare, 1964), potassium nitrate (Solanki and Joshi, 1986), and acidulated water (pH 4.5) with nitric acid (Kumar and Singh, 1987), and it is inhibited by exposing seeds to far-red light (Swarnkar and Kumar, 1977).

Storage Conditions

Radish seeds retain their viability longer than other cruciferous seeds. The economic life of spring and winter radishes is 9 to 12 years (Lazukov, 1969a). Seed germination decreased to 50 percent after 7 years of storage under ambient conditions (Doijode, 1999). Prolonged seed storage increases the percentage of substandard produce and the risk of invasion by many diseases. Bold and well-matured seeds retain better viability than immature seeds (Eguchi and Yamada, 1958). Seed drying to lower moisture helps in better retention of seed viability. Well-dried seeds stored for 9 to 12 years at 10^{-1} to 10^{-2} millibar (mb) preserve high seed quality without loss of viability (Dressler, 1979). Kretschmer and Waldhor (1997a) stored low-moisture seeds (4.2 percent) in plastic bags at $-20°C$ for ten years. There was little change in moisture content, and initial germination (86 percent) remained unchanged. Likewise, seeds with 6 percent moisture packed in polyethylene bags and laminated aluminum foil pouches and stored at 5 and $-20°C$ maintained viability for 15 years. The retention of viability during storage is better at $-20°C$ than at $5°C$. These stored seeds when sown in the field did not show any morphological abnormalities for root and shoot characteristics (Doijode, 1999).

Storage fungi. Storage fungi such as *Alternaria alternata, Aspergillus flavus,* and *Fusarium moniliforme* associate with seeds during storage, especially under high levels of seed moisture or relative humidity. Seed treatment with captan, thiram, or carbendazim controls the fungi.

Invigoration of Stored Seeds

Seed viability is lost gradually during storage. However, loss of vigor precedes the decrease in viability. Viability can be revived to a certain extent by exposing stored seeds to physical or chemical stimuli. Seed soaking in water followed by drying improves germination and seedling vigor (Doijode and Raturi, 1990). Soaking leaches out certain toxic substances and activates the enzymes involved in germination. Savino, Haigh, and Leo (1979) attributed this to the counteraction of free radicals that damage cellular components.

CARROT: *Daucus carota* L.

Introduction

Carrot is a popular root crop in temperate, tropical, and subtropical regions. It is cultivated as a fresh produce item and also processed, canned, and dehydrated. Tender carrots are used in soups, stews, curries, and confectioneries and cooked as a vegetable. Carrot juice is a popular drink in developed countries. Carrots are rich in carbohydrates, fiber, minerals, especially iron, and vitamins, such as alpha- and beta-carotene. Carrot roots, particularly black cultivars, are fed to cattle. Carrot seed contains an essential oil that acts as a carminative and stimulant and has wide usage in the treatment of kidney ailments.

Origin and Distribution

Carrot genetic diversity exists in Africa, Europe, and West Asia. It probably originated in Afghanistan. It is cultivated in almost all countries. Some leading producers are Argentina, Brazil, Canada, China, France, India, Indonesia, Italy, Malaysia, Mexico, Morocco, the Netherlands, Nigeria, the Philippines, Poland, Russia, the United Kingdom, and the United States.

Morphology

Carrot belongs to family Umbelliferae and contains chromosome number 2n = 18. It is a dicotyledonous biennial plant with a short stem and well-developed roots. Its swollen roots, which are either short and blunt or long with a tapering cone, are edible. It is cultivated as an annual vegetable and as a biennial for seed production. The stem elongates and gives inflorescence during the second year. The inflorescence is a compound umbel that is flat or round. Flowers open from the outer orbit and proceed toward the center. Flowers are perfect and white, with five stamens and an inferior ovary with two locules. Fruit is oblong-ovoid, ciliate, and indehiscent and contains a single seed, which contains essential oil. One thousand seeds weigh about 0.8 g.

Seed Storage

Carrot is cultivated by seeds. Seeds are also used in crop improvement and in conserving genetic diversity through seed storage. Seeds show orthodox storage behavior; they can withstand desiccation and chilling during prolonged storage. Well-dried seeds retain viability for longer periods at low storage temperatures. Seeds with less than 10 percent moisture remain viable for six months under ambient conditions (Agrawal, 1980). Further, sealed storage preserves seeds for eight years (Dutt and Thakurta, 1956).

Seed Collections

Carrot requires a cooler climate for growth and development. The Asian types produce seeds in plains, while European types require vernalization of six to eight weeks at 8°C for induction of flowering. However, seeds are normally produced in hills, having higher altitude and a cooler climate. Carrot grows well in deep, sandy loam, well-drained soil. Seeds are produced either by seed-to-seed or root-to-seed method. The latter is a widely accepted practice, requiring the selection of good roots for planting. Plants are spaced properly, adequately fertilized, timely irrigated, and protected against pests and diseases. Planting full-size roots, with one-half or two-thirds top, gives higher seed yield (Lal and Pandey, 1986). A closer spacing of 45 × 20 cm also gave higher seed yield (Malik, Singh, and Yadav, 1983). Flowering occurs at eight-day intervals in different orders of umbel (see Figure 37.1). Individual flowers are protoandrous, and insects, especially bees, bring about cross-pollination. An isolation distance of 1,600 m is maintained between varieties. Off-type plants showing variation for root and shoot characteristics are removed. The primary and secondary umbels are selected for seed purposes, as they give good-quality seeds. Seeds from primary umbels are heavier than those from secondary ones (Gray, Steckel, and Ward, 1983). The crop is harvested when the third-order umbels turn brown. Plants are cut at the base and stored for curing. Subsequently, seeds are separated, cleaned, dried, and packed for storage.

Seed Germination

Fresh carrot seeds show a considerable amount of dormancy (Brocklehurst and Dearman, 1980), which is attributed to underdeveloped embryos and immature seeds (Gray, 1979). Mature seeds also show dormancy

FIGURE 37.1. Carrot Flowering in Different Orders of Umbel

(Watanabe, Asano, and Maeda, 1955). According to Doust and Doust (1982), an after-ripening period of three months eliminates seed dormancy. Comparatively, wild relatives of carrots show more pronounced dormancy than cultivated ones. Seed germination is higher at a constant temperature from 15 to 25°C or at an alternate temperature of 20/30°C for 16/8 h, respectively. Seed germination is greater in vigorous seeds, which complete the germination process in a short time. Seeds with low vigor take longer for seedling emergence and exhibit greater variability in plant characteristics. Likewise, seed germination is lower and delayed in immature as compared to mature seeds (Thomas, Gray, and Biddington, 1978).

Seed size. Seed size also affects seed germination. Malik and Kanwar (1969) reported that field emergence is greater for large seeds. Likewise, large seeds (1.66 mm) have greater storability. The crop yield is also increased by 40 percent when large as compared to smaller seeds are sown (Usik, 1980).

Chemicals. Seed germination is improved by soaking seeds in naphthaleneacetic acid (100 ppm) (Bhat, 1963). Seed soaking in water exudates certain chemicals, including toxic ones. According to Chaturvedi and Murallia (1975), these exudates contain a high amount of inhibitors that affect the germination process. The amount of exudates mainly composed of water-soluble sugars is negatively correlated with seed viability (Dadlani and Agrawal, 1983).

Seed priming. Seed priming promotes germination in fresh and old seeds. Seeds soaked in polyethylene glycol (6000) at 15°C for ten days resulted in early and higher germination (Yanmaz, 1994).

Storage Conditions

Low-temperature storage. Seeds attain full viability and vigor on maturity. Such seeds retain their viability for a longer period than immature seeds (Eguchi and Yamada, 1958). Seed germination and seedling vigor decrease when seeds are stored at 20°C and 80 percent RH. Seeds lose viability rapidly at higher temperatures and relative humidity. The ideal storage conditions for longer storage life are temperature close to 0°C and 70 percent RH. Loss of seed viability is more rapid in dry seeds than in ultradry seeds at 20°C (Ellis et al., 1996). Seeds with 2.0 to 3.7 percent mc (ultradry) and those with 5.5 to 6.8 percent mc (dry) retained viability for five years at –20°C, while seeds with 7 percent moisture exhibited 60 percent germination in unsealed, and 80 percent in sealed, conditions after two years of ambient storage (Doijode, 1995). Seeds stored in moisture-proof containers at room temperature remained viable for 22 months. Seed stored in polyethylene bags maintained higher viability than those stored in paper bags. Seed viability was maintained for eight years at 5 and –20°C storage (Doijode, unpublished data). Likewise,

Kretschmer and Waldhor (1997b) reported that seeds stored in plastic bags maintained viability for 11 years at –20°C storage.

Storage with desiccants. Desiccants are effective in lowering seed moisture. Seeds stored with silica gel retained viability for 36 months under ambient conditions (Varier and Agrawal, 1989). Horky (1991) reported that desiccants are ineffective at a constant storage temperature of 0°C and they improve seed storage quality in the presence of fluctuating temperatures between 5 and 30°C.

Modified-atmosphere storage. A higher level of oxygen in the storage container affects seed viability. Several volatile compounds, such as methanol, ethanol, acetone, and acetyldehyde, are released from dry seeds during storage. The quantity of volatile compounds released increases with increasing storage period and temperature (Zhang et al., 1993). Lowering the oxygen level and/or replacement with inert gases, such as nitrogen or carbon dioxide, maintains high germination during ambient storage.

Storage of pelleted seeds. Storage of pelleted seeds enhanced germination by 4 to 8 percent compared to nonpelleted seeds. Further, the yield from pelleted seeds is greater by 10 percent (Konstantinov and Petkov, 1982).

Storage fungi. Storage fungi are largely associated with improper storage, especially when high seed moisture is retained during ambient storage. Fungi are effectively controlled by seed treatment with thiram, which also promotes seed germination (Miller and Linn, 1957). Elaine and Shiel (1980) suggested coating seeds with resins and incorporating fungicides during storage to protect seeds from fungal attack.

Invigoration of Stored Seeds

Soaking stored carrot seeds in water or in dilute chemicals, such as sodium thiosulfate (10^{-5} M) or disodium phosphate (10^{-4} M), for 2 h followed by drying reduces physiological deterioration during subsequent storage. Even hydration by saturated atmosphere for 24 to 48 h followed by drying is effective in reducing loss of seed viability (Kundu and Basu, 1981). Midstorage hydration-dehydration is effective in improving seed viability and vigor (Pan and Basu, 1985).

BEETROOT: *Beta vulgaris* L.

Introduction

Beetroot is an important salad crop in temperate regions. It is also known as garden beet. Here, the edible portion includes the swollen hypocotyl and a smaller part of the taproot. The foliage is also used as greens and cooked as a vegetable. Apart from its usage in salad, beetroot is boiled, pickled, or cooked as a vegetable. It is a good source of carbohydrates, calcium, phosphorus, potassium, and vitamin C. Its foliage is rich in iron and vitamin A.

Origin and Distribution

Beetroot originated in Asia Minor, the Mediterranean region, and southern Europe. It is a popular vegetable crop in Europe, the Middle East, and North America. It is cultivated in Canada, China, Denmark, Germany, India, Indonesia, Italy, Malaysia, the Netherlands, the Philippines, the United Kingdom, and the United States.

Morphology

Beetroot is a biennial glabrous herb that is cultivated as an annual for vegetable purposes. It belongs to family Chenopodiaceae and has diploid chromosome number $2n = 18$. The food is stored in the root, which is flat, conical, or tapering; red; and swollen. The red pigment is due to the presence of betacyanins. Foliage is dark green or red and ovate, forming a rosette. Inflorescence is a three- to four-flowered cyme; it grows up to 1 m in height. Flower is small, with a green calyx, no corolla, and five stamens, with one ovule per ovary. Fruit is an aggregate formed by the partial fusion of two or more fruits. In this, the calyx continues to grow after fertilization; becomes hard, uneven, and corky; and encloses the seed. It is referred to as beet seed or glomerule and normally contains two to six seeds. True seeds are small, brown, 3 mm long, and kidney shaped. One thousand seeds weigh about 17 g.

Seed Storage

Beetroot seeds show orthodox storage behavior. The retention of high seed quality, viability, and vigor is the desired end in seed storage. Seeds with high initial vigor maintain viability for a longer period. Seeds are viable for two to six years under normal storage conditions (Nath, 1976).

Seed Collections

Beetroot is a cool-season crop, requiring a cool and humid climate for optimum growth. It needs deep, fertile, well-drained, sandy loam soils. Heavy soils are unsuitable for cultivation, and it is sensitive to waterlogging. Beetroot is propagated by seeds. Plants require vernalization for induction of flowering, and an altitude over 1,000 m is congenial for seed production. Seeds are commonly produced by the root-to-seed method, which allows root inspection during transplanting. Sometimes the seed-to-seed method is also used. The first healthy and true-to-type stecklings are selected and stored for a short period before replanting. Proper fertilizer dosage and irrigation, especially during the dry period, are applied to the seed plots. Plants are protected from pests and diseases. Beetroot is cross-pollinated, predominantly by wind, so an isolation distance of 1,600 m is maintained between cultivars for maintenance of genetic purity. Off-type plants are removed based on foliage and root characteristics. Fruits become brown on maturity (see Figure 37.2). Plants are cut when the fruits on lower branches are mature, then they are heaped and stored for a short period. Later, the dried fruits are threshed; seeds are separated, cleaned, dried, and suitably packed for storage.

Seed Germination

Seed germination is slow and erratic and takes longer to complete. Fruit surface contains an inhibitor that affects the germination. Simple washing or soaking in water or in plant extracts of certain seaweeds, which appear to possess cytokinin-like substances, improves seed germination (Wilczek and Ng, 1982). Khan et al. (1983) reported that osmoconditioning of seeds with 1.2 MPa of PEG (6000) followed by a two-minute water rinse is effective in improving germination.

FIGURE 37.2. Beetroot Fruits on Maturity

Storage Conditions

Beetroot seeds remain viable for a fairly long period under room temperatures. However, they deteriorate rapidly at higher storage temperatures and humidity. Lowering of seed moisture is beneficial for prolonging storage life and maintaining high seed quality. Seed moisture is reduced by 1 percent through the addition of sodium sulfate at the rate of 1.5 percent of seed weight basis (Erofeev and Zelenin, 1968). Seed storage at lower temperature reduces the metabolism and slows the process of seed deterioration. Seeds stored at 10°C and 50 percent RH maintained high seed viability for 13 years (Maude and Bambridge, 1985).

TURNIP: *Brassica rapa* L.

Introduction

Turnip is an important cruciferous vegetable crop in temperate regions that is, of late, becoming popular in tropical regions as well. It is cultivated for its fleshy roots and foliage. Tender roots are used in salad and are cooked as a vegetable. It is a good source of iron and vitamins A and C.

Origin and Distribution

Turnip originated in the Mediterranean region and is cultivated throughout the world. It is grown in Afghanistan, Canada, Denmark, France, Germany, India, Italy, Japan, Malaysia, the Philippines, Russia, the United Kingdom, and the United States.

Morphology

Turnip belongs to the Cruciferae family, and its diploid chromosome number is 2n = 20. It is a biennial dicotyledonous herb that is cultivated as an annual for vegetable purposes. Food is stored in the taproot, which is flat, globose, or a long globe. Its color above ground is red, purple, white, or yellow, and the underground portion is white or yellow. Foliage is light green and hairy, with lobed leaves. Flower is perfect and bright yellow, with five sepals, five petals, five stamens, and a three-celled superior ovary. Fruit is a dehiscent silique; seeds are round; black, red, or brown; and 1.5 to 2 mm in diameter. One thousand seeds weigh roughly 4.3 g.

Seed Storage

Seeds are used in the commercial production of turnip. They are also employed in evolving new genotypes and in maintaining genetic diversity for longer periods. Seeds show orthodox storage behavior and remain viable for longer periods under dry, cool storage conditions.

Seed Collections

Turnip is a hardy cool-season crop that can withstand frost. It grows well in deep, fertile, well-drained, sandy loam soils. Seeds are produced either by seed-to-seed or root-to-seed method. The latter is commonly used and provides easy selection of desirable types of roots for high seed quality. Healthy true-to-type roots are transplanted. The inflorescence is a terminal raceme. It is a cross-pollinated crop, and honeybees are the main pollinators. It is easily crossable with Chinese cabbage, radish, and mustard. Thus, an isolation distance of 1,600 m is provided to safeguard against genetic contamination. Roguing is done at different times to remove off-type plants. Normal and timely cultural practices, such as fertilizer application, irrigation, and suitable plant protection measures, are followed during seed production. Pods turn yellow on maturity, and the whole flowering stalk is cut, heaped, and allowed to cure for four to five days. Pods are threshed, and seeds are separated, cleaned, dried, and packed for storage.

Seed Germination

Seed germination is epigeal and higher at a constant temperature of 20°C or at alternate temperatures of 20/30°C for 16/8 h. Seed size varies from 1 to 1.7 mm; large to medium-sized seeds give early and higher germination than smaller seeds (Sandhu, Kang, Dhesi, 1964).

Storage Conditions

Turnip seeds store well under ambient conditions. Higher storage temperatures and relative humidity are injurious to seed longevity. Lazukov (1969b) reported that seeds stored at 18 to 22°C and 60 to 80 percent RH remained viable for 12 to 13 years. Economic seed longevity is about two to three times less than biological longevity. Seed moisture is lowered and kept at a safe limit through the use of chemical desiccants. According to Nakamura (1958), calcium chloride is a satisfactory desiccant for longer seed storage. Another desiccant, sodium sulfate, applied at the rate of 1.5 percent of seed weight reduces the seed moisture by 1 percent and keeps seed viable for a longer period. It also protects seeds from pathogens (Erofeev and Zelenin, 1968). Resins are used in storage to prevent premature germination and to facilitate storage by extending storage life (Elaine and Shiel, 1980).

REFERENCES

Agrawal, P.K. 1980. Relative storability of seeds of ten species under ambient conditions. *Seed Res.* 8:94-99.

Bhat, S.K. 1963. The effect of seed treatment in carrot by NAA. *Sci. Cult.* 29:409.

Brocklehurst, P.A. and Dearman, J. 1980. The germination of carrot (*Daucus carota* L.) seeds harvested on two dates: A physiological and biochemical study. *J. Exp. Bot.* 31:1719-1725.

Chaturvedi, S.N. and Murallia R.N. 1975. Germination inhibitors in some umbellifer seeds. *Ann. Bot.* 39:1125-1129.

Dadlani, M. and Agrawal, P.K. 1983. Factors influencing leaching of sugars and electrolytes from carrot and okra seeds. *Scientia Hort.* 19:39-44.

Doijode, S.D. 1995. Effect of packaging and storage temperatures on longevity of carrot seeds. *Indian J. Plant Genetic Resources* 8:233-236.

Doijode, S.D. 1999. Problems and prospects in conservation of valuable genetic resources in radish (*Raphanus sativus* L.). *Indian J. Plant Genetic Resources* 12:100-102.

Doijode, S.D. and Raturi, G.B. 1990. Effect of hydration-dehydration on the storability of tomato and radish seeds. *Indian J. Pl. Physiol.* 33:172-174.

Doust, J.L. and Doust, L.L. 1982. Life history patterns in British Umbelliferae: A review. *Bot. J. Linn. Soc.* 85:179-194.

Dressler, O. 1979. Storage of well dried seeds under vacuum—A new method for long term storage of seeds. *Die Gartenbauwissenschaft* 44:15-21.

Dutt, B.K. and Thakurta, A.G. 1956. Viability of vegetable seeds in storage. *Bose Res. Inst. Calcutta Tran.* 19:27-36.

Eguchi, T. and Yamada, H. 1958. Studies on the effect of maturity on longevity in vegetable seeds. *Natl. Inst. Agri. Sci. Bull. Ser. E. Hort.* 7:145-165.

Elaine, M.S. and Shiel, R.S. 1980. Coating seeds with polyvinyl resins. *J. Hort. Sci.* 55:371-373.

Ellis, R.H., Hong T.D., Astley, D., Pinnegar, A.E., and Kraak, H.L. 1996. Survival of dry and ultradry seeds of carrot, groundnut, lettuce, oilseed rape and onion during five years of hermetic storage at two temperatures. *Seed Sci. Technol.* 24:347-358.

Erofeev, A.A. and Zelenin, V.M. 1968. On the problem of drying and storing vegetable seeds with sodium sulphate. *Trudy permsk selhoz Inst.* 39:428-435.

Gray, D. 1979. The germination response to temperature of carrot seeds from different umbels and times of harvest of the seed crop. *Seed Sci. Technol.* 7:169-178.

Gray, D., Steckel, R.A.J., and Ward, J.A. 1983. Studies on carrot seed production: Effects of plant density on yield and components of yield. *J. Hort. Sci.* 58:83-90.

Horky, J. 1991. The effect of temperature on the long term storage of dry seeds of some selected vegetables. *Zahradonietvi* 18:29-33.

Khan, A.A., Peck, N.H., Taylor, A.G., and Samimy, C. 1983. Osmoconditioning of beet seeds to improve emergence and yield in cold soil. *Agron. J.* 75:788-794.

Konstantinov, G. and Petkov, M. 1982. Growing carrots from pelleted seeds. *Gradinarska i Lozarska Nauka* 19:78-83.

Kretschmer, M. and Waldhor, O. 1997a. Seed storage at −20°C of *Brassica* species, black radish and radish. *Gemuse Munchen* 33:455-456.

Kretschmer, M and Waldhor, O. 1997b. Seed storage at –20°C of lettuce, dwarf bean and carrot. *Gemuse Munchen* 33:504-555.

Kumar, N. and Singh, V. 1987. Effect of acidulated water on seed germination and seedling growth of *Raphanus sativus*. *Natl. Acad. Sci. Letter* 10:5-7.

Kundu, C. and Basu, R.N. 1981. Hydration-dehydration treatment of stored carrot seeds for the maintenance of vigour, viability and productivity. *Scientia Hort.* 15:117-125.

Lal, S. and Pandey, U.C. 1986. Effect of spacing, root cut treatment and toping on seed production of carrot. *Seed Res.* 14:140-143.

Lazukov, M.I. 1969a. Longevity of cruciferous vegetable seeds and their histochemical characters. *Dokl mosk Sel-hoz. Akad KA. Timirjazeva* No. 148:71-76.

Lazukov, M.I. 1969b. Variability of some properties of cruciferous vegetable root crops grown from old seeds. *Dokl mosk Sel-hoz Akad KA. Timirjazeva* No. 153:119-124.

Lee, S.K. and Nichols, M.A. 1978. Some aspects of seed size and plant spacing on the maturity characteristics of radish. *Acta Hort.* 72:191-199.

Malik, B.S. and Kanwar, J.S. 1969. Effect of seed size and stage of harvest of carrot seeds on the germination, growth and yield of carrot. *Indian J. Agric. Sci.* 39:603-610.

Malik, Y.S., Singh, K.P., and Yadav, P.S. 1983. Effect of spacing and number of umbels on yield and quality of seeds in carrot *(Daucus carota)*. *Seed Res.* 11:63-67.

Maude, R.B. and Bambridge, J.M. 1985. Effects of seed treatments and storage on the incidence of *Phoma betae* and the viability of infected red beet seeds. *Pl. Pathol.* 34:435-437.

Mekenlan, M.R. and Millemsen, R.W. 1975. Germination characteristics of *Raphanus raphanistrum*. I. Laboratory studies. *Bull. Torrey. Bot. Club.* 102:243-252.

Miller, P.M. and Linn, M.B. 1957. Viability of cabbage, carrot, sweet corn and lima bean seeds treated with thiram or copper oxide and stored at three different temperatures. *Plant Dis. Rep.* 41:308-311.

Nakamura, S. 1958. Storage of vegetable seeds. *J. Hort. Assoc. Japan* 27:32-44.

Nath, P. 1976. *Vegetables for the tropical region.* New Delhi: Indian Council of Agriculture Research, p. 109.

Pan, D. and Basu, R.N. 1985. Midstorage and presowing seed treatments for lettuce and carrot. *Scientia Hort.* 25:11-19.

Pandita, V.K. and Randhawa, K.S. 1992. Seed quality in relation to seed size in radish. *Seed Res.* 20:47-48.

Pawar, P.R., Joshi, A.T., and Mahakal, K.G. 1977. Effect of seed treatment with gibberellic acid on germination, growth and yield of radish *(Raphanus sativus* L.). *J. Mah. Agri. Universities* 2:63-64.

Sandhu, K.S., Kang, U.S., and Dhesi, N.S. 1964. Bolder seeds for better turnip yield. *Indian Hort.* 8:30.

Savino, G., Haigh, P.M., and Leo, P.De. 1979. Effect of presoaking upon seed vigor and viability during storage. *Seed Sci. Technol.* 7:57-64.

Singh, K. and Dohare, S.R. 1964. Presowing treatments with naphthaleneacetic acid (NAA) in relation to growth and development of radish. *Punjab Hort. J.* 4:160-164.

Solanki, S.S. and Joshi, R.P. 1986. Methods of increasing seed germination of radish *(Raphanus sativus* L.) and cabbage *(Brassica oleracea). Prog. Hort.* 18:274-276.

Swarnkar, P.L. and Kumar, A. 1977. Observations on the influence of red, far red and blue light on the germination of seeds of the radish *(Raphanus sativus). Comp. Physiol. Ecol.* 2:109-110.

Thomas, T.H., Gray, D., and Biddington, N.L. 1978. The influence of the position of seeds on the mother plant on seed and seedling performance. *Acta Hort.* 83:57-66.

Tokumasu, S. 1971. Effect of dry and wet storage upon dormancy in cruciferous vegetables. *J. Japanese Soc. Hort. Sci.* 40:23-28.

Usik, G.E. 1980. Effect of seed size on emergence and yield in onions and carrot. *Intensifik Ovoshchevod Kishinev Moldavian,* pp. 30-33.

Varier, A. and Agrawal, P.K. 1989. Long term storage of certain vegetable seeds under ambient and reduced moisture conditions. *Seed Res.* 17:153-158.

Vasil'eva, V.J. and Lazukov, M.I. 1970. Characteristics of swelling in radish and cabbage seeds of different ages. *Dokl mosk Sel-hoz. Akad KA. Timirjazeva* No. 158:43-48.

Watanabe, S., Asano, H., and Maeda, T. 1955. On germination of the kintoki carrot seeds. I. Delayed germination. *Kagwa. Agril. Coll. Tech. Bull.* 7:27-30.

Wilczek, C.A. and Ng, T.J. 1982. Promotion of seed germination in table beet by an aqueous seaweed extract. *HortSci.* 17:629-630.

Yanmaz, R. 1994. Effect of presowing PEG (polyethylene glycol) treatments on the germination and emergence rate and time of carrot seeds. *Acta Hort.* 362:229-234.

Zhang, M., Liu, Y., Torii, I., Sasaki, H., and Esashi, Y. 1993. Evolution of volatile compounds by seeds during storage period. *Seed Sci. Technol.* 21:359-373.

– 38 –

Cole Crops

CABBAGE: *Brassica oleracea* var. *capitata* L.

Introduction

Cabbage is an important winter vegetable crop widely cultivated in Europe and North America. It is also popular in the tropics and subtropics. It is cultivated for its tender, modified edible leaves, which form a compact structure called a head. Leaves are used for salad purposes, processed for pickles, sauerkraut, and cooked as a vegetable. Cabbage is a rich source of amino acids, vitamins A and C, and minerals such as phosphorus, calcium, potassium, sodium, and iron.

Origin and Distribution

Cabbage originated in the Mediterranean region. It is grown in China, Colombia, Czechoslovakia, Denmark, Egypt, France, Germany, India, Indonesia, Italy, Japan, Korea, Malaysia, the Netherlands, the Philippines, Poland, Russia, South Africa, Spain, Taiwan, Ukraine, the United Kingdom, and the United States.

Morphology

Cabbage belongs to family Cruciferae and has a chromosome number of 2n = 18. It is a biennial but is cultivated as an annual. The stem is short and encircled by compact foliage. Cabbage heads are round, flat, or pointed. Leaves are red, green, smooth, or wrinkled (see Figure 38.1). Flowering is induced on vernalization. Flowers are small, hermaphrodite, and yellow, with four sepals and petals, six stamens, and an ovary with two locules. Fruit is a silique that dehisces on maturity and contains 12 to 20 seeds. Seeds are small, smooth, brown, and globular. One thousand seeds weigh about 3.3 g.

225

FIGURE 38.1. A Portion of a Cabbage Seed Plot

Seed Storage

Cabbage seeds are short-lived under ambient conditions. They show orthodox storage behavior and can withstand drying as well as chilling temperatures on storage. Seeds are stored for short periods for growing of crops, and for longer periods for the conservation of genetic diversity. A high plant population is a prerequisite for successful crop production, and achieving the required plant density depends on seed quality. Thus, to meet the requirement, suitable storage practices have to be followed for maintaining high viability and vigor during storage.

Seed Collections

Cabbage is a cool-season crop that withstands frost. Hilly regions are generally well suited for seed production because chilling temperatures are prevalent. Normally, temperatures less than 10°C for five or six weeks are required for the induction of flowering. Cabbage grows well in deep, fertile, well-drained soils. Seeds are produced by head-to-seed or seed-to-seed method. In the former method, the crop is raised for head purposes, and well-developed healthy and mature heads are selected for seed purposes and stored for a short period before replanting. Later, these are planted in the field by making a crosscut on top of the heads, without causing injury to the growing points. This facilitates early emergence of flower stalks. The crop is adequately fertilized, irrigated during dry periods, and protected against pests and diseases. In the seed-to-seed method, selected heads are left in the field and crosscuts are made to facilitate flower emergence. Cabbage is a cross-pollinated crop, and bees and flies mainly bring about pollination. Crops are separated by a distance of 1,600 m from another varietal block. Pods turn yellow on maturity, and at this stage, they are collected, dried, and threshed; seeds are separated, cleaned, and suitably packed for storage.

Seed Germination

Seed germination is epigeal and higher at a constant temperature of 20°C. Seed germination in dormant seeds can be improved by exposing to alternate temperatures of 20/30°C for 16/8 h, respectively. Seed color changes to dark brown at maturity but lightens on seed aging. The germination percentage is higher in darker seeds than in lighter ones (Gugnani, Banerjee, and Singh, 1975). According to Liou and colleagues (1989), low germination and emergence rates are associated with low-vigor cabbage seeds, which are improved by soaking seeds in water followed by osmopriming with polyethylene glycol (PEG) (Varier et al., 1995).

Storage of germinated seeds. Germinated cabbage seeds are viable for a short period. Storage life could be extended up to five days if stored in aluminum foil pouches at 5°C. Longer storage reduces seedling emergence. The decrease in vigor is attributed to degradation of structure and enzymic activity (Sunil, 1991). According to Finch-Savage and McQuistan (1988), germinated seeds can be stored for one week at –20°C when cooled at the rate of 2 to 6°C per min.

Storage Conditions

Initial seed vigor, seed moisture, storage temperature, relative humidity, and oxygen content regulate seed storability. Higher seed moisture and higher temperatures promote germination and deterioration. Elaine and Shiel (1980) suggested the use of resins for preventing premature seed germination. Seed coating with resin gives better storage but does not affect the germination process.

Seed moisture. High seed moisture reduces the storage life of seeds. Mackay and Flood (1970) reported that seeds with low initial germination and high moisture content deteriorated faster at 11 to 19°C and 58 to 83 percent RH. Seeds having 3 percent moisture retained their viability for 22 years at –10°C (Ramiro, Perez, and Aguinagalde, 1995). Seeds readily absorb moisture from the atmosphere; thus, they must be packed suitably to protect against high humidity.

Various kinds of packaging, such as glass jars, polyethylene bags, and aluminum foil pouches or cans, are used for seed storage. Normally, low-moisture seeds are sealed in packages before storage. Zink and Camargo (1967) reported that seeds with high moisture content packed in plastic bags deteriorated faster and lost viability after 16 months of storage, whereas low-moisture seeds maintained high viability for 54 months. Seeds attain moisture equilibrium earlier in hessian (burlap) or cloth bags than in polyethylene bags. Lowig (1969) noted that polyvinylidene chloride (PVDC) strengthened paper bags are effective moisture barriers, preserving seed viability for seven years.

Seed storage with desiccants. Seed moisture removal by desiccants is faster than air drying. Desiccants are effective in maintaining high seed viability under fluctuating temperatures (Horky, 1991). Erofeev and Zelenin (1968) recommended that cabbage seeds be mixed with sodium sulfate (1.5 percent) on a seed weight basis for a 1 percent reduction in moisture. With this, high seed viability was preserved for two years, and fungal microflora were reduced on seeds. Further, seed longevity was extended to eight years for desiccant stored seeds as compared to nine months for seeds in open storage (Dutt and Thakurta, 1956).

Temperature. Cabbage seeds lose their viability rapidly at higher storage temperatures. They lost viability completely by four months of storage at 40°C and 77 percent RH (Harrington and Setyati-Harjadi, 1966). Horky (1991) reported that seed deterioration is greater at variable ambient temperatures, but seeds preserve high viability for 86 months at 0°C. Similarly, Lazukov (1969) preserved cabbage seed viability for 7 to 15 years at 18 to 22°C and 60 to 80 percent RH. Low-temperature cabinets, such as a freezer, are effective in preserving seed viability. Accordingly, seeds stored in airtight jars at −20°C preserved their viability for 23 years (Reitan, 1977).

Air. Higher oxygen levels in storage containers increase the respiration rate, resulting in quick deterioration. According to Dressler (1979), cabbage seeds stored in a vacuum at 10^{-1} to 10^{-2} mb maintain viability for 9 to 12 years. Also, high levels of carbon dioxide inhibit the germination process and help in preserving seed viability (Thornton, 1944).

Storage fungi. Cabbage seeds associate with several storage fungi. Seed viability is affected by the presence of *Alternaria, Aspergillus, Fusarium, Rhizopus,* and *Sclerotium* spp. These can be effectively controlled through seed treatment with captan or carbendazim. Miller and Linn (1957) reported that cabbage seeds treated with thiram could be stored for 12 months at 2 to 32°C, with no loss of viability, and treated seeds germinated better than untreated ones.

CAULIFLOWER: *Brassica oleracea* var. *botrytis* L.

Introduction

Cauliflower is an important cole crop cultivated for its edible curd. The curd is made of highly metamorphosed abortive flowers on thick hypertrophied branches. It is cooked as a vegetable and eaten raw. The green foliage serves as good cattle feed. Cauliflower is rich in minerals such as potassium, sodium, iron, phosphorus, calcium, and magnesium. It grows well in temperate regions.

Origin and Distribution

Cauliflower, which evolved from wild cabbage, is native to the Mediterranean region. It is largely cultivated in Australia, Canada, Caribbean islands, China, Egypt, France, Germany, India, Italy, Malaysia, the Netherlands, the Philippines, the United Kingdom, and the United States.

Morphology

Cauliflower belongs to family Cruciferae and has chromosome number $2n = 18$. It is a dicotyledonous biennial plant cultivated as an annual. It has a short, stout stem with long, dark green foliage that encloses the curd, protecting it from sun. The inflorescence is dense and terminal, with a large number of flowers. Flowers are small and have four sepals and petals, six stamens, and a bicarpellary gynoecium. Self-incompatibility exists in certain cultivars. Fruit is a silique about 8 cm long. Seeds are round and dark brown. One thousand seeds weigh about 2.8 g.

Seed Storage

Seeds are widely used in cultivation, and as such, farmers prefer to preserve seeds for shorter periods, such as until next growing season. Seeds are also used in evolving new cultivars and for long-term conservation of genetic diversity. Cauliflower seeds show orthodox storage behavior and withstand desiccation and chilling temperatures during storage. Seeds are short-lived under ambient conditions.

Seed longevity is primarily controlled by genetic factors and further regulated by storage conditions. Doijode (1989) reported that seed storability varies in different cultivars. On deterioration, poor-storer cultivars showed greater leakage of electrolytes and soluble sugars as compared to good-storer ones. Good-storer cultivars preserve viability for relatively longer periods under ambient conditions and do not require any special packaging or conditions for short-term storage.

Seed Collections

Cauliflower is a cool-season crop and is less tolerant to heat and dry conditions. It grows well in cool, moist, fertile, well-drained soils. Seeds are sown in the nursery, and young seedlings are later transplanted in the field. Normal cultural practices, such as timely fertilizer, irrigation, and plant protection measures, are followed. Seeds are produced either by growing in situ or by transplanting selected curds. Normally, plants with good curds are allowed in the field, and later they produce flowers and seeds. These selected plants are scattered all over the field, occupying a large cultivable area. To overcome this, selected plants are transplanted carefully to a seed production block. Curds are scooped during transplanting to facilitate flowering. Cauliflower is highly cross-pollinated, and honeybees and bumblebees are the major pollinators. Thus, an isolation distance of 1,000 m is maintained between varieties. Off-type plants are removed. Pod ripening is nonuniform, and pods on lower branches mature first. These are pooled, cured for a few days, and threshed; seeds are separated, cleaned, dried, and suitably packed for storage.

Seed Germination

Seed germination is epigeal and takes five to seven days to complete. Further, it is enhanced by exposing seeds to 20/30°C for 16/8 h and to light. Verma and Pujari (1977) noted that short-duration exposure of seeds to light increased the germination percentage.

Storage Conditions

Cauliflower seeds are viable for a short period under room temperatures. They remain viable for nine months under open storage (Dutt and Thakurta, 1956). Verma and colleagues (1991) reported that well-dried seeds (5 percent mc) packed in aluminum foil pouches could retain high viability under ambient conditions. Likewise, Doijode (1995) obtained a high germination percentage and high seedling vigor in seeds of good-storer cultivars after five years of storage in ambient conditions. Tronickova (1965) recorded greater loss of viability at 15 to 20°C, while maintaining high viability at –20°C for five years. Seeds removed from –20°C chambers further maintained high viability for three months under normal temperatures. Seeds stored in desiccators remained viable for eight years under ambient conditions (Dutt and Thakurta, 1956), whereas those stored in plastic containers at –20°C, for six years (Kretschmer and Waldhor, 1997).

Storage fungi. Storage fungi affect seed health. These fungi grow rapidly in humid conditions. Fungi such as *Alternaria alternata, Alternaria brassicicola, Aspergillus* spp., *Penicillium* spp., and *Rhizopus stolonifera* associate with seeds (Yadav and Duhan, 1992). These fungi cause discoloration and shriveling, increase storage temperature, and reduce seed viability. Maude and Humpherson-Jones (1980) observed 11 genera on cauliflower seeds, namely, *Alternaria, Aspergillus, Chaetomium, Cladosporium, Curvularia, Drechslera, Fusarium, Gloesporium, Phoma, Rhizopus,* and *Stemphylium.* Seed treatment with fungicides such as thiram or carbendazim is effective in controlling storage pathogens (Kaul, 1972).

KOHLRABI: *Brassica oleracea* var. *gongylodes* L.

Introduction

Kohlrabi is also known as knol khol. It is widely cultivated in temperate regions for its swollen edible stem, and its cultivation is expanding to the tropics and subtropics. Kohlrabi is cooked as a vegetable. It contains high amounts of calcium, magnesium, phosphorus, sodium, sulfur, potassium, and vitamins A and C.

Origin and Distribution

Kohlrabi originated in the Mediterranean region. It is largely cultivated in the Caribbean islands, Denmark, France, Germany, India, Indonesia, Italy, Malaysia, the Netherlands, the Philippines, and the United Kingdom.

Morphology

Kohlrabi is a small, dicotyledonous, biennial herb that is cultivated as an annual. Its stem is small and stout and enlarges just above the ground in a sphere shape. The swollen stem is 5 to 10 cm in diameter and white, green, or purple. Flowers are light yellow in color. Fruit is a silique that dehisces on maturity. Seeds are round, 1.5 to 2.0 mm in diameter, and brown. One thousand seeds weigh about 3.2 g.

Seed Storage

Kohlrabi is propagated by seeds. Seeds are also used in hybridization programs for evolving new cultivars and in long-term conservation of genetic variability. Kohlrabi seeds show orthodox storage behavior. Drying and cold storage do not affect seed viability. Kohlrabi seeds are short-lived under ambient conditions. They require suitable storage conditions for longer maintenance of high-quality seeds.

Seed Collections

Kohlrabi is a cool-season crop. It requires low temperature (10°C) for a week to induce flowering. According to Marrewijk (1976), cold treatment for about 8 to 12 weeks followed by 2 weeks at 10°C initiates flowering. Plants are earthen up or they are protected from frost by covering their swollen stems with soil, lifted and stored separately during frost. Normally, the seed-to-seed method produces seeds. Kohlrabi requires deep, fertile, and well-drained soils. Seeds are sown in the nursery, and seedlings are subsequently transplanted to the main field. Plants are fertilized properly, irrigated, and protected against pests and diseases. Uniform and true-to-type plants are selected for seed purposes, and off-types are eliminated. Kohlrabi is cross-pollinated by bees; thus, an isolation distance of 1,600 m is provided to maintain genetic purity. Mature pods are harvested, stored for a few days, and threshed; seeds are extracted, cleaned, dried, and packed for storage.

Seed Germination

Seed germination is epigeal and completes in a week. It is rapid and highest at a constant temperature of 30°C (Singh, Khurana, and Pandita, 1974).

Storage Conditions

Kohlrabi seeds lose viability and vigor rapidly under ambient conditions. Tronickova (1965) reported that seeds stored at 15 to 20°C lost viability rapidly. In open conditions, seeds remained viable for nine months, and longevity was extended to eight years by preserving seeds in desiccators (Dutt and Thakurta, 1956). Seeds stored at subzero temperatures (−20°C) remained viable for six years (Kretschmer and Waldhor, 1997). Further, seeds maintained viability for three months at normal temperatures after storage at −20°C, with no reduction in germination or seedling vigor (Tronickova, 1965).

REFERENCES

Doijode, S.D. 1989. Some biochemical changes in relation to loss of seed viability in cauliflower germplasm. *Indian J. Plant Genetic Resources* 2:136-139.
Doijode, S.D. 1995. Effect of storage on seed viability and vigor in certain cauliflower cultivars. *Prog. Hort.* 27:175-177.

Dressler, O. 1979. Storage of well-dried seeds under vacuum–A new method for long term storing of seeds. *Deut. Gartenbauwissenschaft* 44:15-21.

Dutt, B.K. and Thakurta, A.G. 1956. Viability of vegetable seeds in storage. *Bose Res. Inst. Calcutta Trans.* 19:27-36.

Elaine, M.S. and Shiel, R.S. 1980. Coating seeds with polyvinyl resins. *J. Hort. Sci.* 51:371-373.

Erofeev, A.A. and Zelenin, V.M. 1968. On the problem of drying and storing vegetable seeds with sodium sulphate. *Trudy permsk selhoz Inst.* 39:428-435.

Finch-Savage, W.E. and McQuistan, C.I. 1988. The potential for newly germinated cabbage seed survival and storage at sub-zero temperature. *Ann. Bot.* 62:509-512.

Gugnani, D., Banerjee, S.K., and Singh, D. 1975. Germination capacity in relation to seed coat color in cabbage and mustard. *Seed Sci. Technol.* 3:575-579.

Harrington, J.F. and Setyati-Harjadi, S. 1966. The longevity of vegetable seeds under high temperature and high humidity. *Proc. 17th Int. Hort. Congr. Md.* 1:627.

Horky, J. 1991. The effect of temperature on the long term storage of dry seeds of some selected vegetables. *Zahradnictvi* 18:29-33.

Kaul, J.L. 1972. Comparative effect of various treatments on the microflora and germinability of cauliflower (*Brassica oleraceae* var. *botrytis*) seeds in storage. *Indian Phytopath.* 25:44-47.

Kretschmer, M. and Waldhor, O. 1997. Seed storage at –20°C of *Brassica* species, black radish and radish. *Gemuse-Munchen* 33:455-456.

Lazukov, M.I. 1969. Longevity of cruciferous vegetable seeds and their histochemical characters. *Dokl. mosk Sel-hoz. Akad. Timirjazeva* No. 148:71-76.

Liou, T.D., Wagenvoort, W.A., Kraak, H.L., and Karssen, C.M. 1989. Aspect of low vigor of cabbage seeds. *J. Agric. Res. China* 38:429-437.

Lowig, E. 1969. Long term storage experiment with *Brassica* seed in protective packing. *Saatgut-Wirtsch* 23:28.

Mackay, D.B. and Flood, R.J. 1970. Investigations in crop seed longevity. I. The viability of *Brassica* seeds stored in permeable and impermeable containers. *J. Natl. Inst. Agril. Bot.* 12:84-99.

Marrewijk, N.P.A. 1976. Artificial cold treatment, gibberellin application and flowering response of kohlrabi (*Brassica oleracea* L. var. *gongylodes* L.). *Scientia Hort.* 4:367-375.

Maude, R.B. and Humpherson-Jones, F.M. 1980. Studies on seed borne phases of dark leaf spot and grey leaf spot on brassicas. *Ann. Appl. Biol.* 95:311-319.

Miller, P.M. and Linn, M.B. 1957. Viability of cabbage, carrot, sweet corn and lima bean seeds treated with thiram or copper oxide and stored at three different temperatures. *Plant Dis. Rep.* 41:308-311.

Ramiro, M.C., Perez, G.F., and Aguinagalde, I. 1995. Effect of different seed storage conditions on germination and isozyme activity in some *Brassica* species. *Ann. Bot.* 75:579-585.

Reitan, A. 1977. Storage of cabbage seeds. *Forskning og Forsok i Landburket* 28:487-495.

Singh, B., Khurana, S.O., and Pandita, M.L. 1974. Chemical weed control in cauliflower, knol khol and turnip. *Haryana J. Hort. Sci.* 3:182-189.

Sunil, G.D.J.L. 1991. Storage of pregerminated vegetable seeds. *Laguna Coll. Tech. Bull.* p. 123.

Thornton, N.C. 1944. Carbon dioxide storage. XII. Germination of seeds in the presence of CO_2. *Boyce Thompson Inst. Contrib.* 13:355-360.

Tronickova, E. 1965. The effect of storage at sub-zero temperature on the viability of kohlrabi and cauliflower seeds. *Rostlinna Vyroba* 11:25-34.

Varier, A., Yaduraju, N.T., Singh, U., and Sharma, S.P. 1995. Field emergence of cabbage seeds as affected by hydro and osmopriming treatments. *Seed Res.* 23:116-117.

Verma, O.P., Jaiswal, R.C., Singh, K., and Gautam, N.C. 1991. Influence of storage conditions on the longevity of tomato and cauliflower seeds. *Veg. Sci.* 18:88-92.

Verma, S.P. and Pujari, M.M. 1977. A note on effect of light and temperature on germination of different vegetable seeds. *Indian J. Hort.* 34:166-168.

Yadav, M.S. and Duhan, J.C. 1992. Effect of seed mycoflora on seed quality of cauliflower (*Brassica oleraceae* var. *botrytis*). *Seed Res.* 20:164-165.

Zink, E. and Camargo, L.De.S. 1967. Studies on the storage of cabbage seed. *Bragantia* 26:53-58.

39

Pod Vegetables

FRENCH BEAN: *Phaseolus vulgaris* L.

Introduction

French bean is widely cultivated in temperate, subtropical, and tropical regions. It is also known as snap bean, kidney bean, common bean, green bean, haricot bean, bush bean, navy bean, pole bean, string bean, and frijol. It is cultivated for its edible tender pods and dry seeds. Fleshy pods and immature seeds are cooked as a vegetable, processed for canning, or frozen. Dry seeds are commonly used as a grain legume. French bean is rich in proteins, carbohydrates, potassium, and iron contents.

Origin and Distribution

French bean originated in Central America. Large genetic diversity among primitive cultivars, landraces, and wild relatives exists in Guatemala, Honduras, and southern Mexico. It is largely cultivated in Brazil, Chile, China, Egypt, England, Ethiopia, France, India, Italy, Kenya, Peru, the Philippines, Spain, Tanzania, Turkey, Uganda, the United States, and Venezuela.

Morphology

French bean belongs to the Leguminosae family and has chromosome number 2n = 22. It is a dicotyledonous, bushy, or twining annual vine. Leaves are dark green, alternate, and trifoliate. Flowers are self-fertilized, axillary, with one raceme, and white, pink, or yellow. Pods are narrow, long, fibrous, fleshy, or stringless and contain about 4 to 12 seeds. Seeds are globular to kidney shaped, and color varies from white, yellow, greenish, pink, red, purple, and brown to black. One thousand seeds weigh roughly about 250 to 600 g.

Seed Storage

French bean is raised through seeds. Seeds are commonly used in hybridization programs for crop improvement and in the gene bank for maintenance of genetic variability for longer periods. French bean seeds show orthodox storage behavior. Seed viability is unaffected by decrease in seed moisture content and storage temperature. Seeds survive for a short period under ambient conditions. Further, seed quality during storage will be affected due to injury by mechanical threshing. Borges, Moraes, and Vieira (1991) reported that about 30 percent of seeds were viable after 30 days, and 18 percent after 90 days, of storage under ambient conditions. James, Bass, and Clark (1967) opined that year of seed productivity influences seed longevity, especially under poor storage conditions, but not under controlled ones. However, seeds remain viable for four years at 21.1°C and 50 percent RH. Well-stored seeds germinate and use their food reserves quickly, while poorly stored seeds use only one-fourth of their reserves before cotyledons drop. Such seedlings grow slowly and mature late, and they also yield less, by 27 to 30 percent (Toole, Toole, and Borthwick, 1965).

Seed storability varies in different genotypes to a limited extent. Moreno-Martinez and colleagues (1994) reported that seed longevity varies from 60 to 180 days in different cultivars when stored at 25°C and 75 to 85 percent RH. Cultivars with white seeds lose seed viability more quickly than non-white-seeded cultivars (Vieira and Fonseca, 1986). According to Hernandez-Livera and colleagues (1990), seeds of cultivars Pinto, Texcoco, and Negro Puebla showed more vigor and viability on storage and were considered to have higher storage potential. These may be used as parents for improving seed longevity in offspring.

Vieira (1966) reported that aging of seeds was accompanied by low germination percentage and an increase in the percentage of abnormal seedlings. Such aged seeds give low yield. As the seed germination decreases, the leakage of amino acids, electrolytes, and soluble sugars increases (Doijode, 1990), and seed starch content decreases, while proteins are unaffected (Alizaga, 1990). Free fatty acids and peroxidase increase with an increase in acidity value during storage (Sawazaki et al., 1985).

Seed Collections

French bean grows well in humid and cooler climatic conditions. High seed quality is obtained from seeds produced in cooler conditions. It grows well in moderate rainfall areas. It is propagated by seeds, which are sown directly in the field in rows. Seeds germinate better when the soil temperature is 30°C, while temperatures greater than 35°C are deleterious to germination. Soils rich in nutrients with good drainage are preferred for seed production. French bean is sensitive to soil moisture stress, and most sensitive to salinity (Cucci et al., 1994). It is a self-fertilized plant; however, an isolation distance of 50 m is maintained between varieties for genetic purity. High temperature during flowering affects fruit set. Likewise, excess rainfall damages the crop and encourages pathogens. Plants are suitably protected by timely spraying of plant protection chemicals. Crops are adequately fertilized and regularly watered. Plants are checked during the flowering and fruiting stages, and off types are removed. Pods turn to yellow on maturity, and delay in harvesting shatters the seeds. Dried pods are collected and threshed, and the seeds are extracted, cleaned, dried, and packed suitably for storage.

Seed maturity. French bean seeds grow rapidly after fertilization. There is a continuous accumulation of dry matter in the seeds. On physiological maturity, seeds attain maximum weight and further inflow of dry matter ceases. Simultaneously, pods change color, becoming yellow, which is a useful sign in identifying seed maturity (Chamma, Marcos-Filho, and Crocomo, 1990). According to Sanhewe and Ellis (1996), longevity of developing seeds continues to increase after mass maturity. Maximum seed quality is attained at the end of the seed-filling phase in bean. The slow desiccation occurring during natural maturation is beneficial for development of high seed quality. Seeds of early harvested pods (60 days after sowing) gave 25 percent germination, and germination increased to 99 percent in seeds extracted from fruit pods 75 to 93 days old (Pimentel and Miranda, 1982).

Seed Germination

Seed germination is rapid and epigeal, and it takes about five days to complete. Germination is higher at a constant temperature of 30°C or at alternate temperatures of 20/30°C for 16/8 h. Carvalho and Oliveira (1978) reported that seeds sown with the hilum in the down position germinated earlier and gave the highest germination percentage. However, Bowers and Hyden (1972) recommended that laying seeds in the flat position is better for higher germination. Bold seeds with higher specific gravity produce healthy and vigorous seedlings and outyielded seeds of lower specific gravity (LeRon and Wyatt, 1977). Similarly, higher seed germination and more vigorous seedlings were obtained from larger and heavier seeds (Doijode, 1984). A direct positive correlation is found to exist between seed size and seedling size in French bean (Ries, 1971). Certain physical stimuli also promote germination. Seed exposure to 40 MHz radio frequency electric fields enhances seed germination percentage by reducing hard seed contents (Nelson, Nutile, and Stetson, 1970).

Storage Conditions

Seed moisture. Seed moisture plays a vital role in seed deterioration. Higher seed moisture promotes rapid metabolism and invasion of pathogens, thereby affecting seed quality. Lowering of seed moisture increases storage life. Zink, Almedia, and Lago (1976) reported that moisture content of 6 to 7 percent maintains high viability for 36 months, compared to less than nine months in seeds having 14.2 percent moisture. In another study, seeds having moisture of 9 to 10 percent stored at 11 to 20°C preserved high seed viability for 16 years (Gvozdeva and Zhukova, 1971). Seeds lose viability rapidly, especially at high moisture levels in sealed storage. Seeds with 12.4 to 20.1 percent mc maintain viability for one year when stored in paper bags (Zink, 1970). When dried carefully, seeds stored at 10^{-1} to 10^{-2} mb retained viability for 9 to 12 years without loss of vigor (Dressler, 1979).

Seed packaging. Seeds are hygroscopic, and they exchange moisture rapidly with the atmosphere, especially at higher temperatures. To keep the moisture relatively near desired levels in seeds, suitable packaging is used. According to Monteiro and Silveira (1982), foam boxes, polyethylene boxes, and tin cans are ideal for seed storage, and seeds remain viable for 24 months. Seeds stored in sealed tin cans remained viable for two years at 15°C and 50 percent RH (Pimentel and Miranda, 1982). Seeds stored in plastic bags at –20°C retain viability for 11 years (Kretschmer and Waldhor, 1997). Seeds having 6.5 percent moisture packed in polyethylene bags and laminated aluminum foil pouches retained viability greater than 85 percent at 5 and –20°C after 15 years of storage, without loss of seedling vigor (Doijode, unpublished data). In humid areas, it is advisable to pack seeds under sealed conditions, even if storage temperature is as low as 5°C (Barton, 1966).

Temperature. Low-temperature storage is beneficial for extending the storage life of seeds. Seeds stored at ambient temperatures gave lower germination percentage and lower seedling vigor than those stored in controlled environments (Fonseca et al., 1980). According to Toole and Toole (1960), French bean seeds are short-lived under ambient temperatures. Seeds remained viable for 24 months under ambient conditions and, thereafter, viability decreased until it was lost completely by the forty-eighth month of storage (Doijode, 1988) (see Figure 39.1). Seedling vigor decreases with an increase in storage period. Seeds stored at higher temperature exhibit poor seed quality and accelerated darkening of seed coat. Hughes and Sandsted (1975) noted that seeds stored at 24°C became dark and had poor germination and increased fatty acid content after one year of storage, while seeds stored at 1°C retained their original color and initial viability. In another study, seeds darkened to a greater extent at 25°C than at

FIGURE 39.1. Seed Viability and Vigor During Ambient Storage of French Bean Seeds

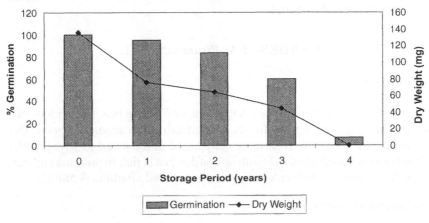

7°C. Low tannin contents are associated with dark seeds (Iaderoza et al., 1989). Seeds stored at room temperature died after two to three years of storage, but those stored at 12°C and 45 percent RH retained high viability for nine years (Almeida and Falivene, 1982). With a rise in temperature and relative humidity, the storage period of seeds decreases. Seeds remain viable for 45 months at 17°C and 35 percent RH, for 30 months at 19°C and 57 percent RH, and for 27 months at 21°C and 57 percent RH (Toole and Toole, 1954). Seeds stored under favorable conditions (19°C and 57 percent RH) gave 23 percent higher germination and higher yield than seeds stored under unfavorable conditions.

Ultralow temperatures are also helpful for maintaining high viability and vigor. However, the utility of this method depends on the regular availability of liquid nitrogen, the size and quantity of seeds to be stored, and cost. Stanwood and Roos (1979) stored French bean seeds in liquid nitrogen (–196°C) for 180 days, with no reduction in germination percentage or vigor.

Storage fungi. High seed moisture or high relative humidity is congenial for the growth and development of storage fungi. Seeds with higher moisture (15 percent) are rapidly invaded by fungi, and their viability is affected. Infestation with *Aspergillus* spp. is more harmful than with *Geotrichum* sp. Seeds with lower moisture content retain their color and viability (Lopez, 1965). Low moisture level in seeds is maintained by using moisture-proof containers, such as tin boxes, which check the growth of microorganisms (Monteiro and Silveira, 1982). Seed deterioration occurs due to toxins produced by fungi that damage the cellular structure (Harman and Granett, 1972). Use of fungicides such as captan, thiram, and carbendazim reduce the damage caused by fungi.

GARDEN PEA: *Pisum sativum* L.

Introduction

Garden pea is also called pea, sugar pea, or Chinese pea. It is widely cultivated in temperate regions for its edible fresh green seeds and pods. The fresh seeds are canned, frozen, or dehydrated and cooked as a vegetable. Dried peas are used as grain legumes. Garden pea is rich in proteins and carbohydrates and contains calcium, phosphorus, and vitamins A and C.

Origin and Distribution

Garden pea originated in Central or Southeast Asia. It is a cool-season crop cultivated in Australia, Belgium, China, Egypt, Ethiopia, France, Hungary, India, Italy, Luxembourg, Myanmar, the Philippines, Russia, Turkey, the United Kingdom, and the United States.

Morphology

Garden pea belongs to the Leguminosae family and has diploid chromosome number 2n = 14. It is an annual dicotyledonous climbing herb with a hollow stem. Leaves are alternate and pinnate with terminal tendrils. Flowers are papilionaceous, hermaphrodite, and axillary, with two or three racemes; have five sepals and five petals; and are white, pink, or purple. Fruit is a pod that varies in size, is straight or curved, and contains two to ten seeds. Seeds are globose, wrinkled or smooth, and green, gray, or brownish. Wrinkled seeds are sweet. One thousand seeds weigh roughly 150 to 330 g.

Seed Storage

Garden pea is propagated by seeds. Seeds are also used in crop improvement and for long-term conservation of genetic diversity. Seeds show orthodox storage behavior. Those with low moisture at low temperature remain viable for longer periods, and when stored under ambient conditions with moisture seldom exceeding 10 percent, they are viable for 18 months (Agrawal, 1980). Optimum moisture content for storage decreases as the storage temperature decreases (Vertucci and Roos, 1993).

Seed longevity is primarily controlled by genetic means and regulated by storage conditions. Kumar and Singhal (1991) observed that dominant maternal effect or some cytoplasmic genes govern the seed longevity.

Seed Collections

Garden pea requires cool, relatively humid climatic conditions for optimum growth. It grows well in fertile, sandy loam soil with good drainage. It is a self-pollinated and long-day plant. An isolation distance of 20 m is maintained between varieties. Crops are provided with adequate fertilizer, irrigation, and plant protection measures for high yield. Seed quality and vigor are improved by application of nitrogen and phosphorus (Hadavizadeh and George, 1989). Plants are checked especially during the flowering and pod development stages for various morphological characters, and off types are removed. Pods turn yellow on seed maturity. The ripe pods are collected and threshed, and the seeds are separated, cleaned, graded, dried, and stored.

Seed Germination

Seed germination is hypogeal and rapid at alternate temperatures of 20/30°C. Fully mature seeds give better germination. Manohar and Sachan (1974) reported that seeds extracted 28 days after anthesis gave higher germination.

Seed size determines the rate of seedling emergence. According to Edwards and Hartwig (1971), small and medium-sized seeds give rapid seedling emergence and greater root development than large seeds, while Kant (1986) noted that large seeds are superior over smaller seeds for early field emergence.

Seed germination is delayed by the presence of hard seeds, as well as by slow imbibition of moisture in certain seeds. This can be overcome by exposing seeds to 40 MHZ radio frequency electric field (Nelson, Nutile, and Stetson, 1970) or by soaking seeds in NAA (30 ppm) and GA (30 ppm) (Doijode and Rao, 1977).

Storage Conditions

Seed moisture. Seed moisture plays an important role in maintenance of seed viability and vigor. Higher moisture is injurious to seed longevity. The optimum moisture for seed storage varies with temperature. Seed moisture of 6 to 7 percent is ideal for longer storage. Pavelkova and Curiova (1984) reported that rapid and excessive removal of moisture affects seed vigor, and also plant yield from such seeds. Drying of immature seeds, especially at higher temperatures, causes hollow heart, a physiological disorder of pea seeds (Perry and Harison, 1973). According to Agrawal and Gupta (1980), hollow heart does not have any adverse effect on pea germination. Dressler (1979) noted that well-dried pea seeds stored at 10^{-1} to 10^{-2} mb maintained viability for 9 to 12 years.

Temperature. Seed viability and vigor are lost rapidly at higher temperatures. Kononkov, Kravchuk, and Vasyanova (1975) reported that seeds failed to germinate during the second month of storage when stored at 20 to 30°C and 82 percent RH, while at a lower temperature, seed deterioration is slow and seeds remain viable for a longer period. Seeds stored at 10°C and 50 percent RH remain viable for five years (James, Bass, and Clark, 1967). Likewise, when stored at subzero temperature (–20°C), seeds remained viable for 33 years (Haferkamp, Smith, and Nilan, 1953).

Storage fungi. Seed health is affected by improper storage conditions. Fungi invade rapidly at higher seed moisture or at high relative humidity. Granett and Harman (1972) reported that seeds infected with *Aspergillus ruber* storage fungi show reduction in seed germination. This was controlled by the use of pentachloro (PCNB) nitrobenzene (Tao, Khan, and Harman, 1974), thiram (Sandhu and Sharma, 1988), and Ceresan and captan (Sohi and Mohalay, 1977).

Seed Deterioration

Changes in seed color and appearance; loss of weight, viability, and vigor; intolerance to unfavorable storage conditions; cellular disintegration; and inactivation of enzymes and DNA synthesis are associated with garden pea seed deterioration. Seed storage under extremely dry (1 percent RH at 10°C) or humid conditions (93 percent RH at 25°C) for six weeks reduced the germination percentage and increased solute leakage (sugar, potassium, electrolytes), and this was attributed to deterioration of the cell membrane (Powell and Matthews, 1977). With an increase in storage period, the electrolyte leakage increases and total phospholipid content decreases (Powell and Matthews, 1981). Seed viability is negatively correlated with the conductivity of leachates (Behal, Hosnedl, and Cinglova, 1982), and it is greater in low-vigor seeds (Nagy and Nagy, 1982). The loss of viability is accompanied by a decline in nucleic acid levels, which includes a reduction in protein hydrolysis, protease activity, catalase activity, and peroxidase activity (Shamsherry and Banerji, 1979). The seed mesocotyl region, including the root and shoot meristem, seems to be the key tissue in pea to protect it from deterioration. This area must remain alive for germination. A few key cells in seeds have to remain functional if the seeds are to retain germinability. These cells are involved with the transportation and assimilation of food (Purkar and Negi, 1982). According to Harman, Khan, and Tao (1976), aging in pea seeds is associated with slower oxygen uptake, a delay in protein synthesis, and reduced enzyme activity. Seed deterioration is slowed by osmotic priming of seeds with polyethylene glycol in acetone (Yang and Ma, 1992) and storage at low moisture and low temperatures.

Invigoration of Stored Seeds

Seed germination improves with presoaking of seeds in water or chemicals, especially in low-quality pea seeds, and this was attributed to the cellular repair mechanism (Savino, Haigh, and Leo, 1979).

BROAD BEAN: *Vicia faba* L.

Introduction

Broad bean is also known by different names, such as favabean, horse bean, Windsor bean, field bean, and tick bean. It is cultivated in temperate, subtropical, and tropical regions for its immature and mature seeds. The seeds are cooked as a vegetable and also frozen or canned. The green crop is fed to cattle and also used as green manuring crop, which fixes the atmospheric nitrogen. Seeds are rich in proteins, carbohydrates, and oil and also contain high amounts of calcium and phosphorus.

Origin and Distribution

Broad bean originated in the Mediterranean region. It is cultivated during the cooler season of the year. It is grown largely in Australia, Brazil, China, Ecuador, Egypt, England, Greece, India, Italy, Mexico, Myanmar, the Netherlands, and Sudan.

Morphology

Broad bean belongs to family Leguminosae and has chromosome number $2n = 12, 14$. It is an erect, dicotyledonous, annual herb that grows up to 1 m in height. Roots are well nodulated. Leaves are alternate and pinnate. The inflorescence is a short axillary raceme. Flowers are white, with purple patches, and fragrant. It is a self-fertilized plant. Fruit is a fleshy pod, 5 to 15 cm long, and contains one to six seeds. Seeds are large, brown, and flattened and vary in shape. One thousand seeds weigh about 800 to 2,000 g.

Seed Storage

Seeds are used in the propagation of broad bean. They are also used in creating and conserving genetic variability. Seeds show orthodox storage behavior. They can withstand loss of moisture without affecting viability and vigor. These seeds also preserve high germination percentage during low-temperature storage.

Seed Collections

Broad bean is a cool-season crop. It grows on high hills in the tropics. It prefers fertile, well-drained soil for cultivation. Seeds are directly sown in the field, and plants are supported during growth and development. It is largely a self-pollinated crop; however, cross-pollination occurs to a certain extent. Thus, an isolation distance of 500 m is provided between varieties to safeguard against genetic contamination. Crops are provided with adequate nutrients, watered regularly, and protected against insect pests and pathogens (see Figure 39.2). Plants are checked during the flowering and fruiting stages, and off-type plants are removed. Pods become brown on maturity. Such pods are harvested when they become relatively dry, then threshed, and seeds are separated, cleaned, dried, and packed for storage.

Seed Germination

Seed germination is hypogeal and higher at a constant temperature of 25 to 30°C or at alternate temperatures of 20/30°C for 16/8 hours. Seeds take longer to germinate due to the presence of a hard seed coat. Germination can be improved by making a small incision on the seed coat or through scarification (Jha et al., 1987). Singh and Singh (1990) reported that seeds

FIGURE 39.2. Broad Bean Vine with Many Mature Pods

soaked in concentrated sulfuric acid for 30 min followed by washing with tap water enhanced the germination percentage. Longer acidification causes seedling injury. Moist stratification at 5°C also improves field emergence and promotes vegetative growth and flowering (Zaki, Helal, and Gabal, 1982).

Storage Conditions

Broad bean seeds are short-lived under ambient conditions. Seed deterioration is greater at high seed moisture and storage temperature. It is associated with loss of viability and reductions in root and shoot growth (Abdalla and Roberts, 1969). At higher relative humidity, seeds gain moisture and lose viability rapidly, especially at higher temperatures. Seeds gave 94 percent germination when stored at room temperature and 76 percent RH, and zero at 30°C and 76 percent RH, after six months of storage, while viability was lost completely at 40°C and 76 percent RH after one month of storage (Lanteri, 1983). Seeds stored in open conditions maintain high viability, particularly at high temperatures. Jute and paper bags are ideal for storing high-moisture seeds. Seeds with moisture content greater than 19 percent showed 86 percent germination after six months of storage (Fuciman, 1989). Seed stored in liquid nitrogen (–196°C) preserve their initial viability for three years, while seed storage at –100 to –120°C results in low germination as compared to storage at –196°C. Fast thawing (10°C/second [sec]) is more effective than slow thawing (1°C/sec) in retrieval of high viability after ultralow-temperature storage of seeds (Fedosenko, 1978).

HYACINTH BEAN: *Lablab niger* Medik.

Introduction

Hyacinth bean is also called dolichos bean, Egyptian bean, Indian bean, or lablab bean. It is cultivated for its immature seeds and pods. Tender seeds and pods are cooked as a vegetable, while mature and dried seeds are used as pulses. Ripe and sprouted seeds are also cooked as a vegetable. Seeds are rich in protein and carbohydrates.

Origin and Distribution

Hyacinth bean originated in India. It is widely cultivated in dry and low-rainfall regions. It is a popular vegetable crop in Brazil, China, Egypt, India, Indonesia, Malaysia, Papua New Guinea, the Philippines, and Sudan.

Morphology

Hyacinth bean is a vigorous, bushy or twining, dicotyledonous, perennial herb. However, it is cultivated as an annual. It belongs to the family Leguminosae and has chromosome number 2n = 22, 24. Leaves are dark green, alternate, and trifoliate. Inflorescence is axillary, and flowers emerge in clusters of four or five. Flowers are hermaphrodite; white, red, or purple; and self-pollinated. Pod is long, oblong, and somewhat curved and contains three to five seeds. Seeds are round and of various colors, such as white, red, brown, black, or variously speckled, with a long white hilum. One thousand seeds weigh about 330 g.

Seed Storage

Seeds are used for propagation and also for long-term conservation of genetic diversity. Seeds show orthodox storage behavior; they can withstand drying and maintain high viability under low-temperature storage conditions.

Seed Collections

Hyacinth bean is a warm-season crop. It grows well in high-temperature and low-rainfall areas. It requires deep, fertile, well-drained soil and is susceptible to waterlogged conditions. Seeds propagate it. Seeds are sown directly in the field, and later, the plants are supported with poles. Crops are properly fertilized, irrigated, and protected against pests and diseases. Although hyacinth bean is a self-pollinated crop, insects bring about cross-pollination to a certain extent. To prevent contamination, a seed plot is isolated from another variety by a 50 m distance. Plants are regularly checked, especially during the flowering and pod development stages, and off-type plants are removed. Pods become light yellow on maturity. It takes about 70 to 120 days from sowing to harvesting, depending upon the cultivar and growing conditions. Ripe pods are harvested, and seeds are separated, cleaned, dried, and suitably packed.

Seed Germination

Seed germination is epigeal and completes in a week. There is no problem of dormancy. Seed germination is higher at a constant temperature of 25°C or at alternate temperatures of 20/30°C for 16/8 h. These seeds are sensitive to moisture stress during germination (Manohar and Mathur, 1975a).

Storage Conditions

Seeds are viable for two years under ambient conditions. Prevalence of high humidity and high temperature hastens the process of seed deterioration and reduces the storage life of seeds. Seeds coated with activated clay (1:100 by weight) retain high viability for 18 months under ambient conditions (Karivartharaju, Palanisamy, and Vanangamudi, 1989). Further, seed packaging in moisture-proof containers, such as aluminum cans or foil, maintains high viability and vigor at low (5°C) or subzero temperatures (−20°C).

Open storage of seeds, especially in high relative humidity areas, leads to the growth of fungi, which discolor seeds, increase temperature, and reduce viability (Prasad and Prasad, 1986). Seed treatment with captan (2 g/kg) preserves high viability and vigor and gives better field emergence after 40 months of storage under ambient conditions (Vanangamudi and Karivartharaju, 1987).

CLUSTER BEAN: *Cyamopsis tetragonoloba* (L.) Taub.

Introduction

Cluster bean is a hardy pod vegetable that is commonly grown in the tropics and subtropics. It is cultivated for its edible pods. Tender pods are cooked as a vegetable. The whole plant is fed to cattle and also used for green manuring purposes. Seeds are rich in protein, carbohydrates, and mannogalactans; the latter have several industrial usages, such as in the textile, paper, and cosmetic industries.

Origin and Distribution

Cluster bean originated in India. Wide genetic diversity for various plant characters, fruit characters, and chemical constituents is found in different parts of India. It is cultivated in Brazil, India, Indonesia, Malaysia, Pakistan, the Philippines, and Sri Lanka.

Morphology

Cluster bean belongs to family Leguminosae and has chromosome number 2n = 14. It is a dicotyledonous, herbaceous, annual, bushy plant. Leaves are medium size, dark green, and trifoliate. Flowers are small, white or pink, and borne on an axillary raceme. Pods are linear, long, compressed, green, erect, beaked, and borne in clusters. Fruit contains 5 to 12 seeds that are 5 mm long; white, gray, or black; and quadrangular in shape. One thousand seeds weigh roughly 25 to 51 g.

Seed Storage

Cluster bean seeds are short-lived under ambient conditions. Seeds are used for commercial cultivation and also as a valuable tool in long-term conservation of genetic diversity. Cluster bean seeds exhibit orthodox storage behavior; they can withstand desiccation and chilling temperatures during storage.

Seed Collections

Cluster bean is a warm-season crop. It grows well in dry conditions and withstands drought. It requires deep, sandy loam soil for successful cultivation. It is grown from the seeds, which are sown directly on ridges in the field. It is a self-pollinated crop, and about 50 m of isolation distance is maintained during seed production. Normal cultural practices, such as spacing, fertilization, irrigation, and plant protection measures, are followed. Short-day conditions are favorable for flowering. The crop is inspected during the flowering and pod development stages, and off-type plants are removed. Seeds mature about 150 days after planting. Pods became light yellow and somewhat hard on maturity. The ripe dry pods are collected and threshed carefully, without injuring the seeds. Seeds are dried, cleaned, and suitably packed for storage.

Seed Germination

Seed germination is rapid and higher at a constant temperature of 30°C and at alternate temperatures of 20/30°C for 16/8 hours. Germination is epigeal and completes within a week. Kalavathy and Vanangamudi (1990) reported that large seeds gave higher germination and seedling vigor than the smaller seeds. During germination, total water-soluble carbohydrates and dry matter contents decrease (Nagpal and Bhatia, 1970). Seed germination is also affected by the use of certain insecticides, such as monocrotophos, on dry seeds (Ramulu and Rao, 1987).

Storage Conditions

Cluster bean seeds survive for a short period at ambient temperatures. Seed quality is affected by improper storage conditions, such as presence of high temperature and relative humidity. Seeds stored in kraft paper bags maintain high viability for 12 months, while those in sealed storage in glass, polyethylene bags, or laminated aluminum foil pouches retain viability relatively longer under ambient conditions (see Figure 39.3) (Doijode, 1989b). Seeds showed 93 percent germination after 15 years of storage at 5 and –20°C. There was a slight reduction in the viability of seeds stored in poly-

FIGURE 39.3. Seed Viability During Open and Sealed Storage of Cluster Bean Seeds Under Ambient Conditions

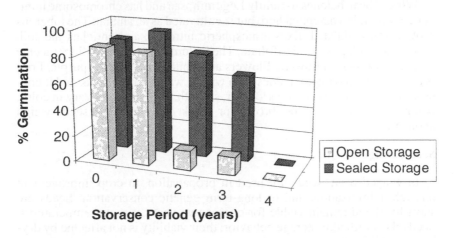

ethylene bags, but no morphological variations were recorded in seedlings raised from these seeds (Doijode, unpublished data). On seed aging, the germination percentage decreases, and the leakage of electrolytes, soluble sugars, and free amino acids increases on seed imbibitions (Doijode, 1989a).

WINGED BEAN: *Psophocarpus tetragonolobus* DC.

Introduction

Winged bean is widely cultivated in humid tropical and subtropical regions. It is also known as four-angled bean, goa bean, Manila bean, and winged pea. Almost all plant parts are used for human consumption. Tender shoots, tuberous roots, young leaves, flowers, immature pods, and mature seeds are cooked as vegetables. Plants are also fed to cattle and used as a green manure. Winged bean is rich in protein, fats, carbohydrates, calcium, phosphorus, and iron.

Origin and Distribution

Winged bean appears to have originated in Southeast Asia, probably in Papua New Guinea. It is cultivated in Ghana, India, Indonesia, Malaysia, Myanmar, Nigeria, Papua New Guinea, the Philippines, Sri Lanka, Tanzania, Thailand, and Vietnam.

Morphology

Winged bean belongs to family Leguminosae and has chromosome number 2n = 26. It is a perennial herb but is cultivated as an annual. The tuberous root is well nodulated, fixes atmospheric nitrogen, and enriches the soil. Leaves are dark green and trifoliate. The inflorescence is an axillary raceme that bears two to ten flowers. Flowers are white, light blue, or purple. Fruit is a green pod, having four sides with wings, which vary in size, and it contains 8 to 17 seeds. Seeds are globular, smooth, and shiny and vary in color, being either white, yellow, brown, or black. One thousand seeds weigh about 500 g.

Seed Storage

In winged bean, seeds are used in propagation, in crop improvement through hybridization, and in long-term genetic conservation. Seeds are short-lived and remain viable for one or two years at room temperatures. Seeds show orthodox storage behavior; their viability is not affected by drying and chilling temperatures during storage.

Seed Collections

Winged bean grows well in hot humid conditions. It requires deep, fertile, well-drained soil. Crops are sensitive to waterlogging. Seeds are directly sown in the field, and the plants are supported by poles. Normal cultural practices, such as fertilization, irrigation, and plant protection measures, are followed. Flowering commences about 50 days after sowing. Winged bean is a self-pollinated crop; however, cross-pollination by bees does occur to a certain extent. An isolation distance of 50 m is maintained between cultivars, and off-type plants are removed. Seeds mature 60 days after anthesis. Pods become brown on maturity. At this stage, pods are collected and dried, and the seeds are extracted.

Seed maturity. Seed maturity determines viability and longevity. Immature seeds exhibit low vigor and viability. Abenoja (1986) noted that seeds harvested in the early days of pod development (5 to 20 days after flowering) showed lowest viability and vigor, compared to late-harvested seeds (25 to 35 days after flowering), which also gave higher germination and better seedling growth. Seeds attain maximum dry weight on the thirtieth day after anthesis.

Seed Germination

Seed germination is hypogeal, and seedlings take longer to emerge from the soil. Hew and Lee (1981) reported that seed germination and seedling growth are better at a constant temperature 25°C and in dark conditions. Seedling emergence is uneven in the field, and this is attributed to the impermeability of the seed coat (Martin, 1978). Uneven emergence was overcome through pregermination softening of the seed coat (Johnson and Madhusudanan, 1989). This includes testa abrasion by sandpaper and acid scarification. Seed soaking in dilute sulfuric acid (8:1 by volume in water) for 5 min at 62°C followed by rinsing in running water for 5 min gave 96 to 100 percent germination, without any injury to emerging seedlings (Csizinszky, 1980).

Storage Conditions

Winged bean seeds lose their viability rapidly at higher temperature and higher relative humidity. Johnson and Madhusudanan (1989) noted that none of the seeds germinated after 24 months of storage under ambient conditions. Seed germination decreased to 50 percent when fresh seeds (20 to 21 percent mc) were stored for two weeks under ambient conditions. Further, seeds lost their viability completely after two months of storage (Hew and Lee, 1981). High-moisture seeds (12.5 to 18.5 percent) stored at 8°C retained seed color, appearance, and high viability for 24 weeks (Onesirosan, 1986).

COWPEA: *Vigna unguiculata* (L.) Walp.

Introduction

Cowpea is cultivated in the tropics for its edible tender long pods and dry seeds. It is also called Bombay cowpea or catjang. Immature pods and fresh seeds are cooked as a vegetable or canned. Green foliage is fed to the cattle. Cowpea is rich in protein, carbohydrates, oils, calcium, phosphorus, and vitamin A.

Origin and Distribution

Cowpea originated in Asia. Wide genetic diversity, including wild relatives, exists in India and China. Major cowpea-producing countries are China, Ghana, India, Indonesia, Malaysia, Nigeria, the Philippines, Sri Lanka, and the United States.

Morphology

Cowpea belongs to family Leguminosae and has chromosome number 2n = 22. It is a dicotyledonous, annual, bushy or climbing herb. Roots contain a large number of nodules. Stems are erect or semierect with trifoliate leaves. The inflorescence is axillary, with two to four flowers in a cluster. Flowers are yellow or violet, perfect, and borne on a long peduncle. Pods are long, thin, white, or green and contain 10 to 30 seeds. Seeds vary in size and shape, from globular to cylindrical. One thousand seeds weigh about 220 g.

Seed Storage

Cowpea is cultivated by seeds. Seeds are used in breeding programs for crop improvement, in long-term genetic conservation, and as a valuable source of proteins. Cowpea seeds show orthodox storage behavior. These seeds tolerate drying, and initial viability can be maintained at low temperatures. Cowpea can survive for only a short period under ambient conditions due to heavy infestation by insects.

Seed Collections

Cowpea is a warm-season crop. It prefers a hot climate for optimum cultivation. It is sensitive to cold and tolerates heat and dry conditions. It grows well in deep, fertile, sandy loam or clay soils with good drainage. Crops are raised by seeds and sown directly in the field. It is self-fertilized plant; however, insects bring about cross-pollination to a certain extent. To safeguard against genetic contamination, an isolation distance of 50 m is maintained between varieties. Crops are provided adequate nutrients, irrigation, and plant protection measures for ideal growth. Plants are checked during the vegetative and reproductive phase, especially during the flowering and fruiting stages, and off-type plants are removed. Pods take about 90 to 150 days to mature, and they become straw colored on maturity. Dry pods must be harvested in time; otherwise, delay leads to shattering of seeds. Pods are threshed, and seeds are separated, cleaned, dried, and packed for storage.

Fruit maturity. Fruit maturity almost coincides with that of seed maturity. Fruit takes about 90 to 150 days to mature, depending upon the cultivar and growing conditions. Cowpea seeds develop rapidly during the earlier period of postfertilization. There is continuous accumulation of dry matter in the seeds, and this ceases on the attainment of mass maturity. Fully matured seeds have high viability and vigor during storage. At this stage, pods change color and become somewhat brittle. The husk development is completed earlier than that of other pod parts. Manohar and Mathur (1975b) reported that seeds attain germination capability in six days after the opening of flowers. However, the germination percentage and seedling vigor are greater for seeds harvested at 14 to 20 days after pollination. Pods could be harvested at the color change stage, with low viability and vigor, to avoid unforeseen unfavorable circumstances, such as bird damage, excess rainfall, and outbreak of other biotic and abiotic stresses. Early harvested seeds (12 days after anthesis) retain viability for six months at room temperature, as compared to more than a year in seeds harvested at 16 to 20 days after flowering.

Seed Germination

Seed germination is epigeal and completes in a week. However, there will be uneven emergence of seedlings due to the presence of hard seeds, which increases with a decrease in water potential. Soil temperature of 25 to 30°C is suitable for better germination. In the laboratory, seed germination hastens at alternate temperatures of 20/30°C for 16/8 h, respectively.

Seed size. In cowpea, seed size varies among and within cultivars, based on seed position in the fruit and growing conditions. Shashidhar, Vyakarnahal, and Swamy (1987) reported that large seeds were more vigorous and gave higher percentage of germination. This was attributed to better food reserves in seeds (Sinha et al., 1988). Large seeds remain viable longer than smaller seeds (Paul and Ramaswamy, 1979).

Storage Conditions

Cowpea seeds are short-lived under ambient conditions. Rodrigo (1935) reported that seeds remain viable for nine to ten months in open storage. Cowpea seeds are easily attacked by insects; thus, a combination of a good physical storage environment and pest-free healthy seeds is required to maintain high seed viability during storage. Durvasula (1990) noted that storage conditions of 20°C and 45 percent RH are ideal for short-term storage of cowpea seeds. Solar heaters are reported to be beneficial in reducing insect activity and in creating dry conditions during ambient storage (Ntoukam et al., 1997).

Seed moisture. High seed moisture is injurious to seed longevity. Seeds with 11.7 percent moisture maintain their viability up to 300 days (Figueiredo, Frazao, Oliveira, and Carvalho, 1982) under ambient conditions. Further, seed drying promotes hard-seededness, and moisture below 6 percent affects germination (Jain et al., 1990). Vanangamudi (1986) reported that low-moisture seeds (8 percent) preserved their viability for 40 months under ambient conditions.

Seed packaging. Seeds are packed to facilitate their handling during storage, and also to create a small storage environment wherein the desired level of moisture or relative humidity can be maintained for a fairly long period. Different types of containers are used according to storage period and storage conditions. Figueiredo, Frazao, Oliveira, and Correa (1982) reported that retention of seed viability was less in paper bags than in cloth and polyethylene bags. Low-moisture seeds are stored in moisture-proof containers such as aluminum foil or cans for better and longer retention of seed quality.

Temperature. High-temperature storage promotes seed deterioration, thereby reducing seed vigor. Seeds lose their viability rapidly at room temperatures and thus store better under refrigeration (Figueiredo, Frazao, Oliveira, and Carvalho, 1982). Insect activity is high in high-moisture seeds at higher temperatures, thereby affecting seed quality.

Storage pests and fungi. Cowpea seeds are severely damaged by pulse beetle (*Callosobruchus chinensis* L.) during storage. Insect activity is rapid in high-moisture seeds, at high temperatures, and in the later period of storage (Figueiredo, Frazao, Oliveira, and Correa, 1982). Application of ethylene dibromide as a fumigant, malathion (5 percent) (Hunje et al., 1990), or neem oil (1:100 parts) (Shivankar et al., 1990) effectively controls pulse beetle infestation. Use of a solar heater during short-term storage of cowpea seeds reduces insect infestation by providing heat sufficient to kill all developing insects inside the seeds (Ntoukam et al., 1997).

The occurrence of heavy rains during harvest increases seed mycoflora and decreases germination percentage. *Aspergillus* spp., which are present externally, and *Fusarium moniliforme* and *macrophomina phaseolina,* which are present internally, affect seed germination. Fungicides such as Ceresan, thiram, carbendazim, and captan are effective in controlling the pathogens and help in maintaining high viability and vigor during storage (Hegde and Hiremath, 1987). Mancozeb and seed coating with red earth (10 g/kg) or activated clay eliminated the storage fungi (Vanangamudi, 1986; Gupta and Singh, 1990).

REFERENCES

Abdalla, F.H. and Roberts, E.H. 1969. The effect of seed storage conditions on the growth and yield of barley, broad beans and peas. *Ann. Bot.* 33:169-184.

Abenoja, R.M. 1986. Seed maturation study in winged bean. *Central Luzon State University Sci. J.* 5:183-184.

Agrawal, P.K. 1980. Relative storability of seeds of ten species under ambient conditions. *Seed Res.* 8:94-99.

Agrawal, V.K. and Gupta, R.K. 1980. Further report of hollow heart in pea seeds. *Seed Res.* 8:75-76.

Alizaga, R. 1990. Physiological and biochemical changes in seeds of three *Phaseolus vulgaris* cultivars with high and low induced vigour. *Agronomia Costarricense* 14:161-167.

Almeida, L.Da.De. and Falivene, S.M.P. 1982. Effects of threshing and storage on conservation of bean seeds. *Revista Brasileira de Sementes* 4:59-67.

Barton, L.V. 1966. The effect of storage conditions on the viability of bean seeds. *Boyce Thompson Inst. Contrib.* 23:281-284.

Behal, J., Hosnedl, V., and Cinglova, O. 1982. Indirect evaluation of the vitality of pea seeds by conductometric testing. *Sbornik Vysoke Skoly zemedelske v Praze Fakulta. Agronomicka* 37:123-138.

Borges, J.W.M., Moraes, E.A., and Vieira, M. 1991. Effect of processing on seed viability of stored beans *(Phaseolus vulgaris)*. *Revista Brasileira de Sementes* 13:135-138.

Bowers, S.A. and Hyden, C.W. 1972. Influence of seed orientation on bean seedling emergence. *Agron. J.* 64:736.

Carvalho, N.M. and Oliveira, O.F. 1978. Effect of seed position at sowing on emergence of *Phaseolus vulgaris* beans. *Cientifica* 6:349-353.

Chamma, H., Marcos-Filho, J., and Crocomo, O. 1990. Maturation of seeds of Aroana beans *(Phaseolus vulgaris)* and its influence on the storage potential. *Seed Sci. Technol.* 18:371-382.

Csizinszky, A.A. 1980. Methods of increasing seed germination of winged bean *Psophocarpus tetragonolobus. HortSci.* 15:252.

Cucci, G., Caro, A.De., Ciciretti, L., and Leoni, B. 1994. Salinity and seed germination of some vegetable crops. *Acta Hort.* No. 362:305-309.

Doijode, S.D. 1984. Effect of seed size on the longevity of seeds in French bean. *Singapore J. Pri. Indust.* 12:62-69.

Doijode, S.D. 1988. Comparisons of storage containers for storage of French bean seeds under ambient conditions. *Seed Res.* 16:245-247.

Doijode, S.D. 1989a. Deteriorative changes in cluster bean seeds in different conditions. *Veg. Sci.* 16:89-92.

Doijode, S.D. 1989b. Effect of temperatures and packaging on longevity of cluster bean seeds. *Die Gartenbauwissenschaft* 54:176-178.

Doijode, S.D. 1990. Influence of temperatures on seed vigour, viability and membrane permeability in different French bean cultivars. *Pl. Physiol. Biochem.* 17:19-22.

Doijode, S.D. and Rao, M.M. 1977. Effect of growth regulators on germination, yield and nutritive value of 'New Line Perfection' peas. *Mysore J. Agric. Sci.* 11:326-329.

Dressler, O. 1979. Storage of well dried seeds under vacuum—A new method for long term storing of seeds. *Deut. Gartenbauwissenschaft* 44:15-21.

Durvasula, R.V. 1990. Interaction effects of environmental conditions and chemical treatment on seed quality of cowpea, *Vigna unguiculata. Laguna Coll. Tech. Bull.,* p. 124.

Edwards, C.J. and Hartwig, E.E. 1971. Effect of seed size upon rate of germination in soybean. *Agron J.* 63:429-430.

Fedosenko, V.A. 1978. The use of super low temp. for long term seed storage (methods and techniques). *Bull. Vsesoyuzno go Ordena Lennia i Ordena Druzhby Narodov Nauchno-Issledovatet Skogo Inst. imeni NI Vavilov* No. 77:53-57.

Figueiredo, F.J.C., Frazao, D.A.C., Oliveira, R.P.De., and Carvalho, J.E.U.De. 1982. Storage of cowpea seeds. *Cir. Tecnica Centro de Pesquisa Agropecuaria do Tropico Umido* No. 31:23.

Figueiredo, F.J.C., Frazao, D.A.C., Oliveira, R.P.De., and Correa, J.R.V. 1982. Storage of cowpea seeds in the physiographic regions of the state of Para. *Cir. Tecnica Centro de Pesquisa Agropecuaria do Tropico Umido* No. 30:48.

Fonseca, J.R., Freire, A.De.B., Freire, M.S., and Zimmerman, F.J.P. 1980. Conservation of bean seeds under three methods of storage. *Revista Brasileira de Sementes* 2:19-27.

Fuciman, L. 1989. Effect of seed storage on growth and development of horse beans. *Sbornik Vysoke Skoly zemedelske v Praze Fakulta Agronomicka Rada A Rostalinna Vyroba* No. 51:197-217.

Granett, A.L. and Harman, G.E. 1972. Seed deterioration due to aging and to infection by *Aspergillus ruber* involves membrane damage. *Phytopathol.* 62:495.

Gupta, A. and Singh, D. 1990. Viability of fungicide treated seeds of mungbean and cowpea in storage. *Seed Res.* 18:70-76.

Gvozdeva, Z.V. and Zhukova, N.V. 1971. The effect of storage conditions on seed longevity in beans, chickpea and soybeans. *Genetike i Selektsii* 45:161-168.

Hadavizadeh, A. and George, R.A.T. 1989. The effect of mother plant nutrition on seed yield and seed vigor in pea *(Pisum sativum)* cultivars. IV. *Inter. Symp. Seed Res. Hort.,* pp. 55-61.

Haferkamp, M.E., Smith, L., and Nilan, R.A. 1953. Relation of age of seed to germination and longevity. *Agron. J.* 45:434-437.

Harman, G.E. and Granett, A.L. 1972. Deterioration of stored pea: Changes in germination, membrane permeability and ultrastructure resulting from infection by *Aspergillus ruber* and from aging. *Physiol. Plant Pathol.* 2:271-277.

Harman, G.E., Khan, A.A., and Tao, K.L. 1976. Physiological changes in the early stages of germination of pea seeds induced by aging and by infection by a storage fungus. *Canadian J. Bot.* 54:39-44.

Hegde, D.G. and Hiremath, R.V. 1987. Seed mycoflora of cowpea and its control by fungicides. *Seed Res.* 15:60-65.

Hernandez-Livera, A.E., Carballo, E., Carballo, A., and Hernandez-Livera, A. 1990. Genetic variation for bean *(Phaseolus vulgaris* L.) seed longevity. *Revista-Chapingo* 15:71-72.

Hew, C.S. and Lee, Y.H. 1981. Germination and short term storage of winged bean seed. *Winged Bean Flyer* 3:15.

Hughes, P.A. and Sandsted, R.F. 1975. Effect of temperature, relative humidity, and light on the color of California light red kidney bean seed during storage. *HortSci.* 10:421-423.

Hunje, R.V., Kulkarni, G.N., Shashidhar, S.D., and Vyakaranahal, B.S. 1990. Effect of insecticides and fungicides treatment on cowpea seed quality. *Seed Res.* 18:90-92.

Iaderoza, M., Sales, A.M., Baldini, V.L.S., Sartori, M.R., and Ferreira, V.L.P. 1989. Polyphenoloxidase activity and changes in color and condensed tanin content in bean *(Phaseolus vulgaris)* cultivars during storage. *Coletanea do Instituto de Tecnologia de Aliments* 19:154-164.

Jain, S.K., Khanna, P.P., Saxena, R.K., Srinivasan, K., Sapra, R.L., and Kapoor, M. 1990. Storage of fumigated cowpea seeds under gene bank conditions. *Inter. Conf. Seed Sci. Technol. New Delhi* Abstr. No. 2.21:43.

James, E., Bass, L.N., and Clark, D.C. 1967. Varietal differences in longevity of vegetable seeds and their response to various storage conditions. *Proc. Am. Soc. Hort. Sci.* 91:521-528.

Jha, B.N., Sinha, S.K., Singh, J.N., and Singh, R.S.P. 1987. Breaking seed dormancy in broad bean. *Seed Res.* 15:226-228.

Johnson, K.M. and Madhusudanan, K.N. 1989. Field emergence, seedling vigor and dormancy in winged bean. *Seed Res.* 17:69-74.

Kalavathy, D. and Vanangamudi, K. 1990. Seed size, seedling vigor and storability in seeds of cluster beans. *Madras Agri. J.* 77:39-40.

Kant, K. 1986. Effect of location and grading on emergence and subsequent plant performance in pea *(Pisum sativum)*. *Seed Res.* 14:102-110.

Karivartharaju, T.V., Palanisamy, V., and Vanangamudi, K. 1989. Influence of seed treatment and storage containers on the viability of lablab *(Lablab purpureus* L.) seeds. *South Indian Hort.* 37:121-122.

Kononkov, P.F., Kravchuk, V.Ya., and Vasyanova, A.V. 1975. The effect of high temp and relative humidity on the changes in seed quality of vegetable crops. *Doklady Vsesoyuznoi Ordena Lenina Akademi Sel. Nauki imeni.* No. 7:15-16.

Kretschmer, M. and Waldhor, O. 1997. Seed storage at −20°C of lettuce, dwarf bean and carrot. *Gemuse-Munchen* 33:504-505.

Kumar, S. and Singhal, N.C. 1991. Effect of aging on parents and their crossed seeds in pea *(Pisum sativum)*. *Seed Res.* 19:51-53.

Lanteri, S. 1983. Effects of unfavorable environmental conditions on broad bean viability. *Sementi Elette* 29:15-19.

LeRon, M.R. and Wyatt, J.E. 1977. Specific gravity of snap bean seeds: Influence on quality attributes. *HortSci.* 12:233.

Lopez, L.C. 1965. Effect of moisture content, microflora and storage time on the viability and external appearance of bean seeds. *Agric. Tech. Mex.* 2:112-115.

Manohar, M.S. and Mathur, M.K. 1975a. Effect of temperature and moisture stresses on germination of seeds. II. Studies on *Dolichos lablab* and *Lycopersicon esculentum*. *Seed Res.* 3:94-101.

Manohar, M.S. and Mathur, M.K. 1975b. Pod development and germination studies on cowpea *(Vigna sinensis)*. *Seed Res.* 3:29-33.

Manohar, M.S. and Sachan, S.C.P. 1974. Pod development and germination studies on pea *(Pisum sativum)*. *Veg. Sci.* 1:22-30.

Martin, F.W. 1978. Observation and experience with winged beans in Puerto Rico. *Winged Bean*, pp. 419-423.

Monteiro, M.R. and Silveira, J.F.Da. 1982. Comparison of containers for storage of bean seeds. *Revista Brasileira de Sementes* 4:47-62.

Moreno-Martinez, E., Vazquez-Badillo, M.E., Navarrete-Maya, R., and Ramirez-Gonzalez, J. 1994. Seed viability of different varieties of bean *(Phaseolus vulgaris)* stored under low and high RH. *Seed Sci. Technol.* 22:195-202.

Nagpal, M.L. and Bhatia, I.S. 1970. Changes in chemical constituents during germination in cluster beans *(Cyamopsis tetragonoloba* Taub.) seed. *Indian J. Agric. Sci.* 40:716-719.

Nagy, J. and Nagy, I. 1982. Relationship between electrical conductivity value and seed vigor in peas. *Novenytermeles* 31:193-205.

Nelson, S.O., Nutile, G.E., and Stetson, L.E. 1970. Effect of radio frequency electrical treatment on germination of vegetable seeds. *J. Am. Soc. Hort. Sci.* 95:359-366.

Ntoukam, G., Kitch, L.W., Shade, R.E., and Murdock, L.L. 1997. A novel method for conserving cowpea germplasm and breeding stocks using solar disinfections. *J. Stored Products Res.* 33:175-179.

Onesirosan, P.T. 1986. Effect of moisture, temperature and storage duration on the level of fungal invasions and germination of winged bean *(Psophocarpus tetragonolobus). Seed Sci. Technol.* 14:355-359.

Paul, S.R. and Ramaswamy, K.R. 1979. Relationship between seed size and seed quality attributes in cowpea *(Vigna sinensis* L.). *Seed Res.* 7:63-70.

Pavelkova, A. and Curiova, S. 1984. Effect of severe dehydration on the biological quality of pea seeds. *Genetika a Slechteni.* 20:251-256.

Perry, D.A. and Harison, G.J. 1973. Causes and development of hollow heart in pea seed. *Ann. Appl. Biol.* 73:95-101.

Pimentel, M.De.L. and Miranda, P. 1982. Physiological maturation and conservation of bean *(Phaseolus vulgaris)* seeds. In *Anais I reuniao nacional de pesquisa de feijao,* Volume 7. pp. 333-335.

Powell, A.A. and Matthews, S. 1977. Deteriorative changes in pea seeds stored in humid or dry conditions. *J. Exp. Bot.* 28:225-234.

Powell, A.A. and Matthews, S. 1981. Association of phospholipid changes with early stages of seed aging. *Ann. Bot.* 47:709-712.

Prasad, B.K. and Prasad, A. 1986. Release of CO_2 from lablab bean *(Dolichos lablab* L.) seeds due to seed borne *Aspergillus niger* during storage. *Seed Res.* 14:250-252.

Purkar, J.K. and Negi, H.C.S. 1982. Initiation of seed deterioration and its localization in peas and wheat. *Seed Res.* 10:196-200.

Ramulu, C.A. and Rao, D. 1987. Effect of monocrotophos on seed germination, growth and leaf chlorophyll content of cluster bean. *Comparative Physiol. Ecology* 12:102-105.

Ries, S.K. 1971. The relationship of size and protein content of bean seeds with growth and yield. *J. Am. Soc. Hort. Sci.* 96:557-560.

Rodrigo, P.A. 1935. Longevity of some farm crop seeds. *Philippine J. Agr.* 6:343-357.

Sandhu, K.S. and Sharma, V.K. 1988. Survival of *Ascochyta* blight of peas through seeds and its control. *J. Res.* 25:53-57.

Sanhewe, A.J. and Ellis, R.H. 1996. Seed development and maturation in *Phaseolus vulgaris* L. II. Post harvest longevity in air dry storage. *J. Exp. Bot.* 47:959-965.

Savino, G., Haigh, P.M., and Leo, P.De. 1979. Effects of presoaking upon seed vigor and viability during storage. *Seed Sci. Technol.* 7:57-64.

Sawazaki, H.E., Teixeira, J.P.F., Moraes, R.M.C., and Bulisani, E.A. 1985. Biochemical and physical modifications of bean seeds during storage. *Bragantia* 44:375-390.

Shamsherry, R. and Banerji, D. 1979. Some biochemical changes accompanying loss of seed viability. *Plant Biochem. J.* 6:54-63.

Shashidhar, S.D., Vyakarnahal, B.S., and Swamy, S.N. 1987. Effect of size grading on seed quality of cowpea. *Seed Res.* 15:214-215.

Shivankar, V.J., Singh, S.N., Khan, A.A., and Tomer, P.S. 1990. Effect of storage conditions on storability of cowpea. *Inter. Conf. Seed Sci. Technol. New Delhi* Abstr. No. 2.31:48.

Singh, H. and Singh, G. 1990. Overcoming the dormancy of broad bean *(Vicia faba)* seeds. *J. Res.* 26:46-48.

Sinha, N.C., Mathur, P.N., Singh, R.P., and Singh, S.N. 1988. Effect of seed size on germination, seed vigor and physiological potential of cowpea. *Seed Res.* 16:41-46.

Sohi, H.S. and Mohalay, M.N. 1977. Effect of different seed dressers on the viability of vegetable seeds. *Indian J. Hort.* 34:199-201.

Stanwood, P.C. and Roos, E.E. 1979. Seed storage of several horticultural species in liquid nitrogen (−196°C). *HortSci.* 14:628-630.

Tao, K.L., Khan, A.A., and Harman, G.E. 1974. Practical significance of the application of chemicals in organic solvents to dry seeds. *J. Am. Soc. Hort. Sci.* 99:217-220.

Toole, E.H. and Toole, V.K. 1954. Relation of storage conditions to germination and to abnormal seedlings of bean. *Proc. Inter. Seed Test. Assoc.* 18:123-129.

Toole, E.H. and Toole V.K. 1960. Viability of stored snap bean seeds as affected by threshing and processing injury. *Tech. Bull U.S. Dept. Agric.* 1213:9.

Toole, E.H., Toole, V.K., and Borthwick, H.A. 1965. Bean storage and yielding ability. *Agric. Res. Wash.* 5:12.

Vanangamudi, K. 1986. Seed storage studies of cowpea. *Trop. Grain Legume Bull.* 33:11-13.

Vanangamudi, K. and Karivartharaju, T.V. 1987. Influence of prestorage seed treatments and storage containers on field emergence and vigor potential of field bean. *Seed Res.* 15:16-19.

Vertucci, C.W. and Roos, E.E. 1993. Theoretical basis of protocols for seed storage. II. The influence of temperature on optimal moisture levels. *Seed Sci. Res.* 3:201-213.

Vieira, C. 1966. Effect of seed age on germination and yield of field bean. *Turrialba* 16:396-398.

Vieira, R.F. and Fonseca, J.R. 1986. Varietal differences in loss of germinability in seeds of *Phaseolus vulgaris*. *Revista Ceres* 33:567-570.

Yang, Y.F. and Ma, Y.S. 1992. Preliminary report on the effect on seed storability of treatment with PEG, acetone and vitamin E. *Crop Genetic Resources* No. 3:36-37.

Zaki, M.E.S., Helal, R.M., and Gabal, M.R. 1982. Effect of vernalization of broad bean seeds on plant vegetative growth, flowering and yielding ability. *Ann. Agric. Sci.* 17:217-226.

Zink, E. 1970. Studies on the storage of snap bean seeds. *Bragantia* 29:107-110.

Zink, E., Almedia, D.A., and Lago, A.A. Do. 1976. Observations on the behavior of bean seeds under different storage conditions. *Bragantia* 35:443-451

40

Leafy Vegetables

FENUGREEK: *Trigonella foenum-graecum* L.

Introduction

Fenugreek is cultivated for food, fodder, and green manure in the tropics and subtropics. Its green leaves are commonly cooked as a vegetable. Seeds contain proteins, carbohydrates, and oils. Seeds are used as spices in culinary purposes and have medicinal properties. Fenugreek is a rich source of vitamins A and C.

Origin and Distribution

Fenugreek is native to Asia and southern Europe. It is cultivated in Egypt, Ethiopia, India, Indonesia, Italy, Malaysia, and Russia.

Morphology

Fenugreek belongs to family Leguminosae and has chromosome number 2n = 16. It is a quick-growing dicotyledonous plant having light to dark green trifoliate leaves. Flowers are small and white. Pods are long, slender, and straw colored and contain 8 to 15 seeds. Seeds are smooth and yellowish brown. One thousand seeds weigh about 10 to 12 g.

Seed Storage

Seeds are commonly used in fenugreek cultivation. They are also used in crop improvement and conservation of genetic diversity. Seeds exhibit orthodox storage behavior; they withstand moisture loss and can be stored at chilling temperatures.

Seed Collections

Fenugreek is a cool-season crop. It grows well in moderate- or low-rainfall areas. It prefers fertile, sandy loam soil. It is propagated by seeds, which are sown directly in the field. Plants are provided with adequate nutrients, irrigation, and plant protection measures. Plants spaced 20 cm apart give higher seed yield than those spaced 40 cm apart (Baswana and Pandita, 1989). It is a self-pollinated crop, and an isolation distance of 50 m is maintained between cultivars. Plants are checked during the flowering and fruiting stages, and off-type plants are removed. Matured pods are collected, dried, and threshed, and the seeds are separated, cleaned, dried, and suitably packed for storage.

Seed Germination

Seed germination is epigeal and higher at a constant temperature of 20°C (Singh, Singh, and Singh, 1990). Fresh seeds exhibit dormancy under natural conditions, even after prolonged storage, due to the presence of a hard, thick seed coat. Seed germination is higher at alternate temperatures of 20/30°C for 16/8 h. Sinha, Jha, and Varshney (1993) reported that seed scarification for 2 or 4 min was effective in reducing the number of hard seeds. Seed germination is affected by saline conditions (Mangal, Yadav, and Singh, 1987) and promoted by gibberellic acid (10 mg·liter^{-1}) application (Tayal and Gopal, 1976).

Storage Conditions

Higher seed moisture and higher storage temperature promote seed deterioration. Well-dried seeds retain their viability for a longer period under cooler conditions. Healthy, bold seeds are selected and dried to 5 to 7 percent moisture for storage. These are suitably packed in polyethylene bags or in aluminum foil pouches and stored at low temperatures for longer storage life.

Storage fungi. Storage fungi grow fast when seed moisture is high, thereby affecting seed quality. Fungi such as *Alternaria, Aspergillus, Cladosporium, Fusarium, Penicillium, Phorma,* and *Rhizoctonia* associate with fenugreek seeds. Seed dressing with captan, carbendazim, and thiram controls the fungi. Boric acid is also effective in eradicating them (Giridhar and Reddy, 1997).

SPINACH: *Spinacia oleracea* L.

Introduction

Spinach is an important winter leafy vegetable cultivated for its smooth or wrinkled edible leaves. It is cooked as a vegetable and used raw in salad. It contains protein, carbohydrates, calcium, iron, and vitamins A, B, and C.

Origin and Distribution

Spinach originated in Central Asia. It is predominantly grown in temperate regions; however, it is also cultivated at higher altitudes in the tropics and subtropics. It is a popular leafy vegetable in Canada, Denmark, Egypt, France, Germany, the Netherlands, Poland, Russia, Spain, the United Kingdom, and the United States.

Morphology

Spinach belongs to family Chenopodiaceae and has chromosome number 2n = 12. It is a dicotyledonous, dioecious, erect, herbaceous biennial plant with edible leaves. The inflorescence is a terminal panicle; male flowers are small and open earlier into pistillate flowers. Female flowers are axillary and greenish and number 6 to 12 per cluster. Fruit is single seeded. Seeds are round and smooth or prickly. One thousand seeds weigh about 10 g.

Seed Storage

Seeds are widely used in crop production. They are also used for evolving new varieties and in the conservation of genetic diversity. Seeds show orthodox storage behavior and are fairly long lived under ambient conditions.

Seed Collections

Spinach is a cool-season crop that grows well in low-temperature areas. High temperature induces premature bolting. Young seedlings can withstand frost. Spinach is a long-day plant. Flowering is profuse under long-day and chilling conditions. The plant bears staminate, pistillate, or hermaphrodite flowers separately. It grows well in fertile, well-drained, sandy loam soil. Crops are fertilized, irrigated, and sprayed regularly against insect pests and diseases. Stems elongate and form flowers in clusters. Spinach is cross-pollinated by the wind; therefore, an isolation distance of 1,600 m is provided between varieties. Roguing is done during the flowering and fruiting stages, and off-type plants are removed. On maturity, plants are cut and threshed, and the seeds are separated, cleaned, dried, and packed for storage.

Seed Germination

Spinach seeds show a considerable amount of dormancy. Prevalence of low temperature also induces dormancy. Such dormant seeds require after-ripening as well as higher temperatures for germination. Khalilov (1977) reported that 21 days of after-ripening was beneficial for eliminating dormancy. Seeds germinate at 5 to 27°C, and the optimum temperature is between 20 and 25°C (Khalilov, 1977; Goyal, Singh, and Singh, 1978). Spinach seeds can also tolerate a high level of soil salinity (Cucci et al., 1994). Seed germination is hastened by exposing seeds to a 40 MHZ radio frequency electric field (Nelson, Nutile, and Stetson, 1970).

Storage Conditions

Seed storage under improper conditions affects seed quality. Seed viability is lost rapidly in high-moisture seeds and also at high storage temperature and relative humidity. Nath (1976) reported that spinach seeds remain viable for three to four years under proper storage conditions. Seeds having 5 to 7 percent moisture that are suitably packed in polyethylene bags and in aluminum foil pouches or cans and stored at lower temperatures maintain high viability and vigor during storage.

AMARANTH (*Amaranthus* spp.)

Introduction

Amaranth is a popular leafy vegetable of the tropics. Different species that are cultivated predominantly as vegetables are *Amaranthus cruentus* L. in Africa, *Amaranthus tricolor* L. in Asia, and *Amaranthus dubius* Mart. ex Thellung in the Caribbean (Grubben, 1977). In the Americas, the varieties of *Amaranthus hypochondriacus, A. cruentus,* and *A. caudatus* are largely cultivated for grain purposes. Amaranth is cultivated for its edible leaves, which are cooked as a vegetable. Also, a few cultivars are grown exclusively for grain purposes, and these grains are used for food purposes. Amaranth is rich in proteins, magnesium, phosphorus, sodium, calcium, iron, sulfur, potassium, and vitamins A and C. It also contains a high amount of oxalic acid.

Origin and Distribution

The origin of amaranth is controversial. Genetic diversity exists in both the New World and the Old World, predominantly in South America and South Asia. Amaranth as a leafy vegetable is mainly grown in Brazil, China, Ethiopia, India, Indonesia, Kenya, Malaysia, Mexico, Myanmar, the Philippines, Taiwan, Uganda, and the United States.

Morphology

Amaranth belongs to family Amaranthaceae and has chromosome number 2n = 32. It is a dicotyledonous, monoecious, short-lived annual plant. Stems are straight, grow up to 1 m in height, and end with terminal inflorescences. Leaves are opposite and green or purple in color. The inflorescence is an axillary or terminal racemose that is green, orange, or purple and contains numerous staminate and pistillate flowers. Flowers have four to five petals, one to five stamens, and two or three carpels with superior ovaries. Fruit is an indehiscent utricle surrounded by a persistent perianth. Seeds are small, black, brown, or white and shiny. One thousand seeds weigh about 0.3 g.

Seed Storage

Seeds are widely used for cultivation, for crop improvement, and in the long-term conservation of genetic diversity. Amaranth seeds show orthodox storage behavior. They remain viable for a short period under ambient conditions. Seed storability increases with drying and storage at low temperatures.

Seed Collections

Amaranth is a warm-season crop and needs warmer conditions for optimum growth. It grows well in hot, sunny places. Seeds are directly sown in the field. Light soil with good drainage is ideal for cultivation. Often the terminal growing point is cut to encourage side shoots. The crop is wind pollinated, but selfing is more common. An isolation distance of 500 m is maintained. Roguing is done to remove off-type plants. Rajan and Joseph (1990) reported that seed quality was unaffected by cutting the leaves for vegetable purposes. Spikes are collected when glumes turn brown and seeds become black. Subsequently, spikes are dried and threshed, and the seeds are separated. Seeds are dried and cleaned, and the good seeds are selected for storage.

Seed Germination

Amaranth seeds show a certain amount of dormancy. Seed storage for short duration eliminates dormancy. Seed germination rate decreases due to low soil temperature. Temperature of 30°C is optimum for better seedling emergence. The process of seed germination is enhanced by exposure of seeds to alternate temperatures (20/30°C for 16/8 h). Aufhammer and colleagues (1998) noted that light or short illumination inhibits the germination process.

Storage Conditions

Seed viability is affected by improper storage conditions. Seed viability is reduced after one year of ambient storage (Aufhammer et al., 1998). Seed viability is preserved by packing seeds in polyethylene bags and in aluminum foil pouches and storing them at 5 and −20°C for ten years (Doijode, unpublished data). Storage of seeds in nitrogen maintains higher germination and seedling vigor for seven years (Doijode, unpublished data).

CORIANDER: *Coriandrum sativum* L.

Introduction

Coriander is cultivated as spinach and mainly used in flavoring food dishes. Dry seeds are powdered and used as a spice in food preparation. It is rich in proteins, carbohydrates, calcium, iron, and vitamin A. Fruit contains a volatile aromatic oil that is used in medicines and has carminative properties.

Origin and Distribution

Coriander originated in the Mediterranean and Middle East regions. It is cultivated in Brazil, China, England, France, Germany, India, Italy, Malaysia, Morocco, Thailand, and the United States.

Morphology

Coriander belongs to family Umbelliferae and has chromosome number 2n = 22. It is an annual, dicotyledonous herb that grows up to 80 cm in height. The stem is hollow, with broad leaves at the base and narrow leaves at the top. The inflorescence is a terminal umbel with numerous small white, pink, or lavender flowers. The fruit is globular, yellow, or brown; ribbed; and about 4 mm in diameter. It is a two-seeded schizocarp. Seeds are semiglobular and contain essential oil (0.1 to 1 percent), mainly coriandrol ($C_{10}H_{18}O$). One thousand seeds weigh roughly 7 to 13 g.

Seed Storage

Coriander is propagated by seeds. It is also used in breeding programs and in long-term conservation of germplasm. Seeds show orthodox storage behavior. Seed viability is unaffected by drying and low-temperature storage.

Seed Collections

Coriander needs a cooler climate for optimum growth, especially at higher altitudes. It grows well in deep, fertile, well-drained soils. Seeds are sown directly in the field. Normal cultural practices are followed for high crop production. It is a cross-pollinated crop, and about 1,000 m of isolation distance is maintained to prevent genetic contamination. Off-type plants are removed. Fruit requires about 90 to 120 days from sowing to maturity. Dry, mature fruits are collected, separated, and cleaned for storage.

Seed Germination

Fresh seeds do not germinate readily, revealing dormancy. Seed germination is better at 15°C, and it takes 10 to 21 days to complete (Jethani, 1982). Seed germination decreases above 30°C and lower than 10°C (Putievsky, 1980). Seeds selected from the second-order umbel gave higher germination and higher seedling vigor than those from the first- or third-order umbels (Ponnuswamy and Ramakrishnan, 1985). Germination is improved by subjecting seeds to alternate temperatures of 27/22°C for 8/16 h, respectively (Putievsky, 1980).

Storage Conditions

Seed viability and vigor decrease with an increase in seed moisture and storage temperatures. Low-moisture seeds are packed in polyethylene bags or in laminated aluminum foil pouches and are stored at low temperature. This maintains high viability for a longer period.

Storage pests and fungi. Higher seed moisture is congenial for insect and pathogen activity. Verma (1988) reported that fumigation with methyl bromide (64 ml·liter^{-1}) or hydrogen cyanide (HCN) (48 mg·liter^{-1}) gas for 2 h at reduced pressure kills the insects inside and does not affect seed quality. Fungal damage is controlled by application of fungicides, such as brassicol, zineb, captan, and thiram, at 0.3 to 0.4 percent (Prasad, 1988).

REFERENCES

Aufhammer, W., Czuczorova, D., Kaul, H.P., and Kruse, M. 1998. Germination of grain amaranth effects of seed quality temperature, light and pesticides. *European J. Agron.* 8:127-135.
Baswana, K.S. and Pandita, M.L. 1989. Effect of time of sowing and row spacing on seed yield of fenugreek. *Seed Res.* 17:109-112.
Cucci, G., Caro, A.De., Ciciretti, L., and Leoni, B. 1994. Salinity and seed germination of some vegetable crops. *Acta Hort.* No. 362:305-309.
Giridhar, P. and Reddy, S.M. 1997. Effect of some food preservative on seed mycoflora of fenugreek. *Seed Res.* 25:92-93.
Goyal, R.D., Singh, M.B., and Singh, P.V. 1978. Enhancement of germination of the seeds of spinach *(Spinacia oleracea). Seed Res.* 6:145-150.
Grubben, G.J.H. 1977. *Tropical vegetables and their genetic resources.* Rome: International Board for Plant Genetic Resources.

Jethani, I. 1982. Revised studies on the seed testing procedures of coriander (*Coriandrum sativum* L.). *Seed Res.* 10:143-149.

Khalilov, M.Kh. 1977. Biology of germination of *Spinacia turkestanica* seeds. *Rastitel'nye Resursy* 13:518-520.

Mangal, J.L., Yadav, A., and Singh, G.P. 1987. Effects of different levels of soil salinity on seed production of leafy vegetables. *J. Res.* 17:47-51.

Nath, P. 1976. *Vegetables for the tropical region.* New Delhi: Indian Council of Agricultural Research, pp. 1-109.

Nelson, S.O., Nutile, G.E., and Stetson, L.E. 1970. Effect of radio frequency electrical treatment on germination of vegetable seeds. *J. Am. Soc. Hort. Sci.* 95: 359-366.

Ponnuswamy, A.S. and Ramakrishnan, V. 1985. Germination and vigor of coriander (*Coriandrum sativum* L.) seeds in relation to their position in umbels of different orders. *South Indian Hort.* 33:30-34.

Prasad, B.K. 1988. Performance of fungicides in storage of coriander seeds at high RH. *Seed Res.* 16:123-125.

Putievsky, E. 1980. Germination studies with seeds of caraway, coriander and dill. *Seed Sci. Technol.* 8:245-254.

Rajan, S. and Joseph, M. 1990. Influence of vegetable harvest on seed yield and quality in amaranthus *(Amaranthus tricolor). Inter. Conf. Seed Sci. Technol. New Delhi* Abstr. No. 1.39:20.

Singh, T., Singh, P.V., and Singh, K.K. 1990. Evaluation of best temperature and medium for testing germination of fenugreek seeds in laboratory. *Seed Res.* 18:31-33.

Sinha, S.K., Jha, B.N., and Varshney, S.K. 1993. Effect of various treatments on hardseededness in Kasuri methi (*Trigonella corinculata* L.). *Seed Res.* 21:114-116.

Tayal, M.S. and Gopal, R. 1976. Synergistic and antagonistic behavior of malic hydrazide, morphactin and GA with reference to the seed germination in fenugreek (*Trigonella foenum-graecum* L.). *Indian J. Pl. Physiol.* 19:71-75.

Verma, B.R. 1988. Effect of multiple fumigation on seed germination. *Seed Res.* 16:241-244.

– 41–

Salad Vegetables

LETTUCE: *Lactuca sativa* L.

Introduction

Lettuce is a popular salad crop in temperate regions. It is a cool-season vegetable crop exclusively cultivated for its edible leaves. It is also grown to a lesser extent and on a smaller scale in the subtropics, where it is cooked as a vegetable. It is rich in minerals such as calcium, phosphorus, and iron and vitamins A and C.

Origin and Distribution

Lettuce originated in the Middle East region composed of Iran and Turkey, where the wild species are found. It might have evolved from *Lactuca serriola* L. Lettuce is cultivated in Australia, Canada, China, France, Italy, Malaysia, the Netherlands, the Philippines, South Africa, Spain, the United Kingdom, and the United States.

Seed Storage

Lettuce seeds show orthodox storage behavior. Seed viability is unaffected by loss of moisture and by exposing seeds to chilling temperatures. Seeds are commonly used in commercial propagation and for long-term genetic conservation.

Morphology

Lettuce belongs to family Compositae and has chromosome number 2n = 18. It is cultivated as an annual for vegetable purposes and as a biennial for seed purposes. Lettuce is a dicotyledonous herb with a small stem that contains latex. Leaves are alternate, sessile, and spirally arranged in a rosette. Outer leaves are dark green and rough, whereas the inner ones are light green and succulent. Heads vary in size and shape and are classified as cos, butter head, or crisp and leafy type. Inflorescence is a capitulum that contains yellow perfect flowers arranged in a cluster. Fruit is a single-seeded achene, black, brown, or white in color. One thousand seeds weigh about 0.6 to 1.0 g.

Seed Collections

Lettuce is a cool-season crop. It grows well under temperate conditions and requires humus and well-drained soil. Dry conditions or water shortage induces premature bolting. It is propagated by seeds. Crops are provided with optimum dosages of nutrients, regular irrigation, and plant protection measures for the production of high-quality seeds. Lettuce is mostly a self-pollinated crop; however, cross-pollination occurs to a small extent. An isolation distance of 50 m is maintained between varieties to safeguard against genetic contamination. To facilitate flowering, a slight cut is made on the head or the upper leaves are removed. Off-type plants are removed during the flowering and fruiting stages. Fruits take two weeks to mature after anthesis and are harvested when the capitulum bursts and shows the pappus. Seeds shatter if harvesting is delayed. Plants are cut and placed in small piles, dried, and threshed. Seeds are separated, cleaned, and packed for storage.

Seed Germination

Lettuce seeds germinate readily and do not show any dormancy. However, seeds exposed to conditions of extreme temperature or light show secondary dormancy. The optimum temperature for seed germination is 25°C, and germination takes four or five days to complete. Seeds do not germinate when exposed to higher temperatures. Damania (1986) reported that a temperature regime greater than 25°C inhibits the germination process, due to the impermeability of the seed coat to oxygen. Continuous exposure of seeds to a dark period also induces secondary dormancy (skotodormancy) (Hsiao, Vidaver, and Quick, 1984).

Light. Light plays a vital role in lettuce seed germination. It promotes germination activity under a set of conditions. Sarma and Chakraborthy (1977) reported that germination rate is higher in the presence of light than in darkness. The exposure of moist seeds to fluorescent light for 24 h alone and/or followed by 24 h of darkness accelerates the germination process (Verma et al., 1974). The uneven emergence of seedlings caused by higher temperature is overcome by exposing moist seeds to red light for 3 h (Kretschmer, 1982, 1983; Hsiao, 1993), and this eliminates the secondary seed dormancy (Hsiao, Vidaver, and Quick, 1984). Globerson (1981) reported that while red light improves germination, far-red light inhibits it.

Chemicals. Seed germination improves by imposing certain chemical treatments, such as osmoconditioning with PEG at 10 bar (Kretschmer, 1982), acid immersion for an hour followed by washing (Hsiao, 1993), and application of ethephon (Verma, Bohra, and Sankhla, 1973), gibberellic acid (50 mg·liter^{-1}) or kinetin (Pauli and Harriott, 1968), thiourea (1,000 mg·liter^{-1}) (Sarma and Chakraborthy, 1977), and vitamin K (50 ppm) (Mullick and Chatterji, 1978). Germination is affected by volatile toxic substances released within storage containers (Kummer, 1953).

Hsiao, Vidaver, and Quick (1984) reported that immersion of skotodormant seeds in acid followed by either gibberellic acid application or exposure to red light improved germination. Acid treatment weakens the membrane barrier of the endosperm cells and results in better penetration by gibberellic acid, thereby inducing higher germination.

Storage of pregerminated seeds. Pregerminated seeds are viable for a very short period. However, they can be stored up to five days at 5°C in aluminum foil pouches. Further storage reduces vigor and viability due to degradation of the cellular structure and inactivation of enzymes (Sunil, 1991).

Storage Conditions

Seed moisture or relative humidity, storage temperature, and oxygen play a major role in deterioration of lettuce seeds. Proper and careful regulation of these parameters is beneficial in retention of high viability on long storage. Lettuce seeds show good viability even after 15 years of storage (Khoroshailov and Zhukov, 1973).

Seed moisture. Seed moisture is reduced with the development of seeds. It remains constant at the physiological maturity of seeds. Further, moisture is reduced by drying seeds to a safe limit for better storage. Lettuce seeds deteriorate rapidly under high humidity and temperatures (Akamine, 1943). Seeds gain or lose their moisture from or to the atmosphere based on changes in vapor pressure. Seeds stored at fluctuating humidities at 5°C or 0 percent RH at room temperature maintain viability for four years (Bacchi, 1960). At highest and lowest humidity, dormancy developed in seeds and was apparently associated with seed deterioration, leading to death. The best seed viability is maintained at 46 to 58 percent RH (Kosar and Thompson, 1957). Kraak and Vos (1987) observed more abnormal seedlings when seed moisture exceeded 12 percent, and seeds were killed within a day when moisture exceeded 47 percent (Stashnov and Dzhantila, 1983).

Excessive drying also affects seed quality. Crop yield decreased in seeds stored at 4.1 percent mc compared to those at 6.4 percent mc (Nowosielska and Schneider, 1980). Dressler (1979) preserved well-dried seeds at 10^{-1} to 10^{-2} mb for 9 to 12 years without loss of viability.

Seed packaging. Packaging creates a barrier between seeds and the external environment. It protects the seeds from high relative humidity and prevents the entry of pathogens. The selection of packaging depends on the duration of storage. Normally, low-moisture seeds are kept in sealed storage. Seeds stored in thinner-gauge bags lose their viability rapidly and exhibit more abnormalities than those stored in thicker-gauge bags (Lowig, 1958). Seeds stored in polyethylene bags at $-20°C$ lost moisture from 6.6 to 5.6 percent in 11 years of storage, and there was no reduction of seed viability during the period (Kretschmer and Waldhor, 1997).

Desiccants. Chemical desiccants are effective in lowering seed moisture, thereby extending storage life. High seed quality was maintained for 86 months when lettuce seeds were stored with desiccants under fluctuating temperatures (5 to 30°C) (Horky, 1991). This method is quite beneficial for short-term storage, easy to manage, and inexpensive.

Temperature. Seed longevity decreases with increasing temperature. Seeds remain viable for four years at 5 or $-18°C$ (Kretschmer, 1976). At 20°C, viability is lost more rapidly in dry seeds (5.5 to 6.8 percent mc) than in ultradry seeds (2.0 to 3.7 percent mc). Ultradry seeds maintained viability for five years at $-20°C$ storage (Ellis et al., 1996).

Air. High oxygen content affects seed quality during storage. Harrison and Mcleish (1954) reported that lettuce seeds retain high viability in sealed containers when stored with carbon dioxide, which maintains higher germination, reduces chromosomal aberrations, and delays aging of seeds. Rao and Roberts (1990) opined that low-moisture seed survival was better in nitrogen than in air, vice versa for high-moisture seeds.

Invigoration of Stored Seeds

Seed quality declines during ambient storage. Midstorage hydration-dehydration of medium-vigor lettuce seeds effectively reduces physiological deterioration and gives better field emergence and higher yields, whereas soaking-drying is not effective in higher-vigor seeds (Pan and Basu, 1985). To improve germination, seeds are to be soaked in water for 2 h followed by storage at 4.4°C for four to six days. Seed soaking in thiourea (0.5 percent) also improves the germination percentage (Dowdles, 1960). The hydration-dehydration treatment curtails the process of deterioration (Pan, Punjabi, and Basu, 1981) and minimizes chromosomal aberrations (Rao and Roberts, 1990). Introduction of chemicals such as ethephon (30 ppm) (Verma, Bohra, and Sankhla, 1973), potassium iodide (10^{-4} M), p-hydroxybenzoic acid (10^{-4} M), or tannic acid (10^{-5} M) into the seeds by dry permeation technique, using acetone as a solvent on the old seeds, reduces seed deterioration and promotes germination by counteracting the effect of free-radical damage to the stored seeds (Basu, Pan, and Punjabi, 1979).

Moist storage. Higher seed moisture causes a rapid decline in longevity and also produces chromosomal aberrations. Seeds stored under an imbibed state do not germinate, which helps in the retrieval of high seed quality. Powell, Leung, and Bewley (1983) reported that imbibed seeds held at 25°C in darkness retained viability up to ten months. Seeds behave differently at a higher level of moisture (15 percent), and oxygen is essential for survival. At this stage, the relative effect of temperature on decreasing longevity is slightly diminished. Seeds at 20 to 30 percent moisture do not suffer further decreases in longevity. It appears that hydration activates the repair mechanism (Ibrahim and Roberts, 1983). Seedlings grown from dry seeds stored under ambient conditions showed an increase in morphological abnormalities with an increase in storage period, whereas seedlings from imbibed stored seeds appeared normal (Villiers, 1974). The repair mechanism operates only under the imbibed state, under the conditions that prevent germination during storage (Villiers and Edgcumbe, 1975).

CELERY: *Apium graveolens* L.

Introduction

Celery is an important salad crop in Europe and North America that is grown for its edible leaf stalks and swollen roots. It is consumed raw or cooked as a vegetable. In temperate regions, the petioles are eaten as a fresh vegetable after blanching or flavoring. Dry leaves are used as a flavoring agent and in medicine. Seeds contain 2 to 3 percent volatile oil. Celery is rich in minerals such as calcium, phosphorus, and iron and in vitamin A.

Origin and Distribution

Celery originated in Asia and Europe. It is largely cultivated in Europe and North America. Some of the major celery-growing countries are Algeria, Canada, China, Egypt, France, Indonesia, Malaysia, New Zealand, the Netherlands, the Philippines, Sweden, the United Kingdom, the United States, and the West Indies.

Morphology

Celery belongs to family Umbelliferae and has chromosome number 2n = 22. It is a dicotyledonous biennial crop. It completes a life cycle in one year under temperate conditions. Celery roots are thick and fleshy; leaves are pinnately compound with long petioles up to 40 cm. Foliage is light or dark green and possesses a strong flavor. The inflorescence consists of small white flowers in an umbel. Flowers are perfect and self-fertile. Fruit is a rigid schizocarp that contains one seed. Seeds are small, brown, and ribbed. One thousand seeds weigh about 0.5 g.

Seed Storage

Celery is propagated by seeds. Seeds are important in genetic conservation and crop improvement. Seeds are moderate storers under ambient conditions. They store well under dry, cool climatic conditions.

Seed Collections

Celery is a cool-season crop that grows well in low-temperature regions. It requires deep, fertile, well-drained, humus soil. It is raised by seeds in the nursery and transplanted in the field. Seeds are soaked in water before sowing for early and higher germination. Plants undergo vernalization for the induction of flowering. Celery is a self-fertile plant, but insects bring about cross-pollination to a limited extent. Thus, an isolation distance of 500 m is provided between varieties. Adequate horticultural practices, such as irrigation, nutrition, and crop protection measures, are followed. Crops are inspected during the vegetative and reproductive phases, and off-types plants are removed. Fruits become brown on maturity, and at this stage, crops are harvested. Delay in harvesting causes shattering of seeds. Later, the plants are threshed, and the seeds are separated, dried, and suitably packed for storage.

Seed Germination

Seed germination is epigeal and higher at alternate temperatures of 15/22°C (Guy, 1981). In celery, the seed embryo is comparatively underdeveloped, and embryo size varies in different cultivars. According to Flemion and Uhlmann (1946), about 20 percent of seeds are empty. Further, immature embryos are dormant (Thompson, 1974). All these factors contribute to delayed and uneven emergence of seedlings. Seed dormancy is also dependent on the umbel order. Thomas, Gray, and Biddington (1987) reported that seeds from primary and secondary umbels were large and less dormant but gave lower percentage of germination than tertiary and quaternary umbels. This is attributed to the imbalance of gibberellin and cytokinin levels in seeds. Dormancy is also induced in seeds by exposing them suddenly to dry conditions after the imbibing state (Biddington, 1981), while seed germination improves with a combination of GA and PEG. Incorporation of growth substances into PEG is more effective than soaking seeds in gibberellins and ethylene solution alone, either before or after osmotic priming (Brocklehurst, Rankin, and Thomas, 1983).

Storage Conditions

Celery seeds maintain high viability and vigor at low temperature and lower humidity. The choice of temperature range and packaging is selected based on duration of storage. For short storage, well-dried seeds can be packed in polyethylene bags and stored at 5°C. For longer storage, seeds can be packed in moisture-proof containers, such as laminated aluminum foil pouches, and then stored at subzero temperatures.

ASPARAGUS: *Asparagus officinalis* L.

Introduction

Asparagus is an important vegetable in Europe. It is cultivated for its edible tender green shoots (spears). Immature shoots are cooked as a vegetable and also processed and canned.

Origin and Distribution

Asparagus originated in the Mediterranean region. It is widely cultivated in Australia, China, England, France, Germany, Hungary, Indonesia, Japan, Kenya, Malaysia, Mexico, the Netherlands, Spain, Taiwan, the United States, and the West Indies.

Morphology

Asparagus belongs to family Liliaceae and has diploid chromosome number 2n = 20, 40. It is a monocotyledonous dioecious herb. It has large fleshy tuberous roots that emerge from an underground rhizome. Leaves are narrow and dark green. Male flowers are small and yellowish green, while female flowers are somewhat large. Fruit is a berry that becomes red on maturation and contains one to three seeds. Seeds are round, flattened, and black in color. One thousand seeds weigh about 20 to 40 g.

Seed Storage

Seeds are used in propagation of asparagus. They are also used in crop improvement. Asparagus seeds show orthodox storage behavior. Higher storage temperatures and relative humidity affect seed longevity.

Seed Collections

Asparagus is a cool-season crop that grows well in rich humus soil. Seeds used for propagation are sown in the nursery and later transplanted to the field. Normal cultural practices are adopted during growing. It is a cross-pollinated crop, and insects mainly bring about pollination. About 1,000 m of isolation distance is maintained between two genotypes. Off-type plants are removed regularly to maintain genetic purity. Ripe berries are harvested, and seeds are extracted, washed, cleaned, dried, and packed for storage.

Seed Germination

Low germination is a common phenomenon in asparagus seeds (Scheer, Ellison, and Johnson, 1960), which show slight dormancy that can be removed by after-ripening for three weeks of storage (Komoti, 1956). However, longer storage, for three months, is better for higher germination (Owen and Pill, 1994). For higher viability and vigor, seeds extracted from red ripe berries are superior over mature bronze berries. Similarly, seeds from a late-harvested crop show low viability.

Seed size. Seed size also determines the rate of germination. Large and medium-sized seeds gave a higher coefficient of germination and better seedling growth (Belletti, 1985).

Storage Conditions

Seed longevity predominantly depends on storage conditions. However, the initial vigor of seeds contributes to a certain extent. Seeds stored under improper conditions, such as high relative humidity and high temperature, deteriorate faster and lose their viability and vigor rapidly. Factors that promote delay of germination during storage enhance the seed longevity. Komoti (1957) reported that drying induced secondary dormancy in seeds. Similarly, a higher level of CO_2 concentration in storage induces secondary dormancy. Seed storage at lower temperature (4°C) maintains higher viability and vigor than storage at a higher temperature (20°C). Well-dried seeds stored at a lower temperature have better storage life.

REFERENCES

Akamine, E.K. 1943. The effect of temperature and humidity on viability of stored seeds in Hawaii. *Hawaii Agri. Exp. Stn. Bull.* 90:23.

Bacchi, O. 1960. Seed storage studies. V. Lettuce. *Bragantia* 19:12-14.

Basu, R.N., Pan, D., and Punjabi, B. 1979. Control of lettuce seed deterioration. *Indian J. Pl. Physiol.* 22:247-253.

Belletti, P. 1985. Correlation of weight and external surface of seed to the percentage and rate of germination in *Asparagus plumisus* var. *nanus*. *Asparagus Res. Newsletter* 3:15.

Biddington, N.L. 1981. Thermodormancy and prevention of desiccation injury in celery seeds. *Ann. Appl. Biol.* 98:558-562.

Brocklehurst, P.A., Rankin, W.E.F., and Thomas, T.H. 1983. Stimulation of celery seed germination and seedling growth with combined ethephon, gibberellins and polyethylene glycol seed treatments. *Pl. Growth Regulation* 1:195-202.

Damania, A.B. 1986. Inhibition of seed germination in lettuce at high temperature. *Seed Res.* 14:177-184.

Dowdles, D. 1960. Germinating lettuce in summer. *Qd. Agric. J.* 86:774.

Dressler, O. 1979. Storage of well dried seeds under vacuum—A new method for long term storing of seeds. *Die Gartenbauwissenschaft* 44:15-21.

Ellis, R.H., Hong, T.D., Astley, D., Pinnegar, A.E., and Kraak, H.L. 1996. Survival of dry and ultradry seeds of carrot groundnut, lettuce, oilseed rape and onion during five years hermetic storage at two temperatures. *Seed Sci. Technol.* 24: 347-385.

Flemion, F. and Uhlmann, G. 1946. Further studies of embryoless seeds in the Umbelliferae. *Boyce Thompson Inst. Contrib.* 14:283-293.

Globerson, D. 1981. Germination and dormancy in immature and fresh mature lettuce seeds. *Ann. Bot.* 48:639-643.

Guy, R. 1981. Influence of de la temperature sur la duree'de germination des semences de especes potageres. *Revue Suisse de Vit.d' Arborl. Hort.* 13:219-225.

Harrison, B.J. and Mcleish, J. 1954. Abnormalities of stored seeds. *Nature* 173: 593-594.

Horky, J. 1991. The effect of temperatures on the long term storage of dry seeds of some selected vegetables. *Zahradnictvi* 18:29-33.

Hsiao, A.I. 1993. Actions of acid immersion, red light and gibberellin A_3 as treatment on germination of thermodormant lettuce seeds. *Environ. Expt. Bot.* 33:397-404.

Hsiao, A.I., Vidaver, W., and Quick, W. 1984. Acidification, growth promoter and red light effects on germination of skotodormant lettuce seeds. *Canadian J. Bot.* 62:1108-1115.

Ibrahim, A.E. and Roberts, E.H. 1983. Viability of lettuce seeds. I. Survival in hermetic storage. *J. Expt. Bot.* 34:620-630.

Khoroshailov, N.G. and Zhukov, N.V. 1973. Long term storage of seed samples in collections. *Trudy po Prikladnoi Botanika Genetike I Selektsii* 49:269-279.

Komoti, S. 1956. Studies on temperature treatment of seeds. I. Effects of temperature treatments on germination of garden asparagus seeds. *Hokkaido Natl. Agril. Exp. Res. Bull.* 70:42-49.

Komoti, S. 1957. Studies on temperature treatments of seeds. II. Dormancy and germinating temperature in garden asparagus seeds. *Hokkaido Natl. Agril. Expt. Res. Bull.* 73:9-19.

Kosar, W.F. and Thompson, R.C. 1957. Influence of storage humidity on dormancy and longevity of lettuce seeds. *Proc. Am. Soc. Hort. Sci.* 70:273-276.

Kraak, H.L. and Vos, J. 1987. Seed viability constant for lettuce. *Ann. Bot.* 59: 343-349.

Kretschmer, M. 1976. Influence of several years storage at different temperature on *Lactuca sativa.* I. Variation of temperature tolerance in darkness and light. *Die Gartenbauwissenschaft* 41:229-235.

Kretschmer, M. 1982. Extension of the temperature tolerance of *Lactuca* achenes with PEG and red light. *Die Gartenbauwissenschaft* 47:152-157.

Kretschmer, M. 1983. Preplanting treatment of lettuce seeds for germination at high temperatures. *Gemuse* 19:48-50.

Kretschmer, M. and Waldhor, O. 1997. Seed storage at –20°C of lettuce, dwarf bean and carrot. *Gemuse-Munchen* 33:504-505.

Kummer, H. 1953. Injury to vegetable seeds stored in color printed seed packets. *Angew Bot.* 27:115-142.

Lowig, E. 1958. Damage to lettuce seed during storage. *Saatgut-Wirtsch* 10: 102-104.

Mullick, P.C. and Chatterji, U.N. 1978. Inhibition of seed germination and early growth of lettuce by joint action of vitamin K and IAA. *Sci. Cult.* 44:177-178.

Nowosielska, B. and Schneider, J. 1980. The productive value of lettuce *(Lactuca sativa)* seed after long term storage. *Hodowla Rosl Aklimat-i-Naslenin* 24:731-739.

Owen, P.L. and Pill, W.J. 1994. Germination of osmotically primed asparagus and tomato seeds after storage up to three months. *J. Am. Soc. Hort. Sci.* 119:636-641.

Pan, D. and Basu, R.N. 1985. Midstorage and presowing seed treatment for lettuce and carrot. *Scientia Hort.* 25:11-19.

Pan, D., Punjabi, B., and Basu, R.N. 1981. A note on the involvement of protein synthesis in hydration-dehydration treatment of lettuce seeds. *Seed Res.* 9:202-205.

Pauli, A.W. and Harriott, B.L. 1968. Lettuce seed selection and treatment for precision planting. *Agri. Eng. St. Joseph Mich.* 49:18-22.

Powell, A.D., Leung, D.W.M., and Bewley, J.D. 1983. Long term storage of dormant Grand Rapids lettuce seeds in the imbibed state: Physiological and metabolic changes. *Planta* 159:182-188.

Rao, N.K. and Roberts, E.H. 1990. The effect of oxygen on seed survival and accumulation of chromosome damage in lettuce *(Lactuca sativa). Seed Sci. Technol.* 18:229-238.

Sarma, C.M. and Chakraborthy, P. 1977. Effect of gibberellic acid and thiourea singly and in combination on the germination of lettuce seeds. *Indian J. Agric. Sci.* 47:18-21.

Scheer, D.F., Ellison, J.H., and Johnson, H. 1960. Effect of fruit maturity on asparagus seed germination. *Proc. Am. Soc. Hort. Sci.* 75:407-410.

Stashnov, S. and Dzhantila, O. 1983. The resistance of hydrated lettuce seeds to low temperature during deep freezing. *Referativnyli Zhurnal* 6:55:373.

Sunil, G.D.J.L. 1991. Storage of pregerminated vegetable seeds. *Laguna Coll. Tech. Bull.,* p. 123.

Thomas, T.H., Gray, D., and Biddington, N.L. 1987. The influence of the position of the seed on the mother plant on seed and seedling performance. *Acta Hort.* 83:57-66.

Thompson, P.A. 1974. Effect of fluctuating temperatures on germination. *J. Expt. Bot.* 25:164-175.

Verma, C.M., Bohra, S.P., and Sankhla, N. 1973. Lettuce seed germination reversal of salinity induced inhibition by ethylene. *Curr. Sci.* 42:294-295.

Verma, S.P., Pujari, M.M., Jain, B.P., and Sinha, A.P. 1974. Effects of interactions of temperature and different duration of fluorescent light and darkness on germination of lettuce seeds. *Proc. Bihar Acad. Agri. Sci.* 22/23:68-71.

Villiers, T.A. 1974. Seed aging chromosome stability and extended viability of seeds stored fully imbibed. *Pl. Physiol.* 53:875-878.

Villiers, T.A. and Edgcumbe, D.J. 1975. On the cause of seed deterioration in dry storage. *Seed Sci. Technol.* 3:761-774.

– 42 –

Cucurbits

CUCUMBER: *Cucumis sativus* L.

Introduction

Cucumber is cultivated in tropical and subtropical regions for its edible fruits. Tender fruits are eaten raw in salad, cooked as a vegetable, and also pickled. They contain carbohydrates and are a rich source of vitamin C.

Origin and Distribution

Cucumber originated in India, and large genetic variability is observed in different parts of India. It is grown in Bulgaria, China, Egypt, Greece, India, Indonesia, Japan, Malaysia, Mexico, the Netherlands, Poland, Russia, Spain, Turkey, and the United States.

Morphology

Cucumber belongs to Cucurbitaceae family and has diploid chromosome number 2n = 14. It is a monoecious, dicotyledonous annual plant. Vines are long and trailing with stout, hairy stems. Leaves are triangular, have three to five lobes, and are cordate. Flowers are yellow; male flowers are borne in clusters and outnumber female flowers. Fruit is a pepo and varies in size and shape, from globular to oblong, and is many seeded. The skin color is whitish green to dark green, turning yellow or brown on maturity. Flesh is pale green with no central cavity. The bitterness of fruit and foliage is owing to the terepene compound cucurbitacin, which gives resistance to certain insect pests. Seeds are white and flat. One thousand seeds weigh 25 to 33 g.

Seed Storage

In cucumber, seeds are used for propagation and in breeding programs. Seeds can be stored until the next growing season or for a very long period for genetic conservation. Cucumber seeds exhibit orthodox storage behavior. They can withstand removal of moisture and can be stored at lower temperatures. High seed viability and vigor are retained at lower temperatures and lower moisture content. Nakamura (1958) noted that cucumber seeds retain their viability for a fairly long period. They lose viability completely by 38 months of storage (Rodrigo, 1953). Germinating power and yield reduce steadily in crops raised from seeds stored under ambient conditions. Crop yield is lower in plants raised from older seeds than in those from fresher seeds (Frohlich and Henkel, 1964). On proper seed storage, low deterioration occurs. Such seeds give more productive plants than fresh ones. Boos (1966) noticed that cucumber plants raised from two- to three-year-old seeds are more productive than plants raised from one-year-old seeds. In fresh seeds, catalase activity is lower than in stored seeds, suggesting that the vigor of freshly harvested seeds is lower than that of stored seeds. Likewise, there was a delay in ethylene production in fresh seed germination (Yin and Cui, 1995). Seed vigor improves in fresh seeds following after-ripening at 75°C for 24 h (Cui and Yin, 1995). Chemical composition of seeds varies with storage period. Sugars and vitamin C tend to increase with longer storage (Zaitseva, 1972).

Seed Collections

Cucumbers are propagated by seeds. High seed viability and vigor are essential for high plant populations in the field and, in turn, for higher crop production. Cucumber grows well in warm climatic conditions. Seed quality improves with growing plants in artificial light, limiting the number of fruits per plant, and pollinating artificially with bees. Seed yield increases and fruit size and number of fruits per plant decrease with an increase in plant population (Cantliffe and Phatak, 1975). Cucumber prefers deep, fertile, well-drained soils. Seeds are directly sown in the field on ridges or on hills. Plants are regularly irrigated, properly fertilized, and protected against pests and diseases. Plants are cross-pollinated mainly by bees, and an isolation distance of 1,600 m is maintained to prevent genetic contamination. Male flowers open earlier than female flowers; the latter are favored by short-day conditions and low temperatures. Plants are to be examined for off types, especially during the flowering and fruiting stages. Ripe fruits are collected when they turn pale yellow or golden in color (see Figure 42.1). Seeds are separated by either mechanical or chemical means. The pulp along with the seeds is fermented. Seeds that sink to the bottom are selected, and floating ones are discarded. Selected seeds are dried, cleaned, and stored.

Fruit maturity. Fruit maturity coincides with seed maturity. Seed maturity determines seed quality for sowing and storage. In cucumber, seed quality is poor in seeds obtained from immature fruits (20 days after anthesis),

FIGURE 42.1. Cucumbers Ready for Harvest

even after subjecting fruits to postharvest ripening. Seeds harvested at 30 days after anthesis need postharvest ripening of 7 to 15 days for better quality. The best seed quality is obtained 40 days after anthesis. Further delay affects seed quality by excessive respiration and field damages (Wallerstein, Goldberg, and Gold, 1981; Nerson, 1991; Nandeesh, Javaregowda, and Ramegowda, 1995). The lipase activity and vitamin C content increase during seed germination. In seedlings raised from seeds extracted from mature fruits, lipase activity, ascorbic acid content, and other biochemical indices were at higher levels than in seedlings raised from seeds of immature fruits (Buriev, 1984). Mature seeds retain viability better than seeds of immature fruits (Eguchi and Yamada, 1958).

Seed Germination

Seed germination is epigeal and completes in five to seven days. It is influenced by fruit maturity at harvest and duration of fermentation, in seed extraction, seed storage time, seed placement, and germination temperatures (Edwards, Lower, and Staub, 1986). Solanki and Seth (1984) reported that a constant temperature of 30°C is ideal for germination. The germination process is completed in the shortest time at higher temperatures. Certain cucumber cultivars germinate more quickly than others, and heritability for speed of germination is very low (0.15 to 0.20 percent) (Wehner, 1982).

Seed dormancy. Shifriss and George (1965) observed dormancy in seeds of cultivated, wild, and hybrid cucumbers that can be overcome by after-ripening for one month (Odland, 1937). Seeds stored for six months at 15 and 20°C showed improved germination. Light inhibits germination, and longer exposure at 20°C induces secondary dormancy.

Seed placement. Seeds placed at 45° angles with the embryos pointed upward give high germination percentage and best seedling growth. This was attributed to better hypocotyl emergence from the soil, rapid cotyledon expansion, and earlier differentiation of leaves (Gomaa, 1980). However, MacNeill and Hall (1982) observed that sowing seeds in the flat position gave higher germination at a constant temperature of 26.7°C.

Chemicals. Seed germination improves with the application of certain growth substances. Seed treatment with GA $_{4/7}$ plus ethephon or insulin or insulin-like factors I and II enhances the germination of cucumber seeds (Staub, Wehner, and Tolla, 1989). Further, seed quality is improved by treating seeds with succinic acid (Boos, 1966).

Storage of pregerminated seeds. Pregerminated seeds can be stored for ten days in a cold-air medium. Seed viability and vigor decrease on longer storage, which is attributed to the degradation of functional structures, an increase in fatty acids content, and enzymatic degradation (Sunil and Mabesa, 1991).

Storage Conditions

Seed viability decreases rapidly at higher storage temperature and relative humidity. Cucumber seeds fail to germinate at 20 to 30°C and 82 percent RH after three months of storage. For shorter seed storage, the optimum conditions are 25°C and 45 percent RH (Kononkov, Kravchuk, and Vasyanova, 1975). Zaitseva (1972) reported that seeds stored at a constant temperature of 16 or 25°C or at varying temperatures of −7 and 20°C maintain viability for seven years. Seed viability decreases rapidly at 20°C and 80 percent RH. The optimum conditions for storage are temperatures close to zero and relative humidity not exceeding 70 percent (Kurdina, 1966). At lower temperature and lower humidity, such as 3°C and 38 percent RH, seeds retain 80 percent germination after ten years of storage. Further, none of the seeds germinated after 13 years of storage (Ali et al., 1991). Seed moisture should be low for ultralow-temperature storage. Seeds with 5 or 6 percent moisture stored at −196°C retained germination ability for two years, and germination was affected when seed moisture increased to 9 to 10 percent (Fedosenko and Yuldasheva, 1976). The safe moisture level is 1.02 to 4.08 percent for ultradry storage. Seeds with this moisture range retain high viability and vigor; those with moisture below this range are damaged by extreme drying, and those with moisture above this range showed reduced storability (Ji, Guo, and Ye, 1996).

Seed packaging. Various kinds of packaging materials are employed in seed storage. The ideal material should protect seeds from high moisture, withstand low temperatures, and preserve viability for longer periods. Sealed airtight containers are better for storage than unsealed containers (Rodrigo, 1953). Cooper (1959) noted that sealed polyethylene film of 0.125 and 0.25 mm thickness is an excellent moisture-resistant package for dry cucumber seeds. Thin polyethylene film of 0.037 to 0.05 mm was useful for a short-term marketing package, but not for long-term storage.

Storage fungi. Seed quality is affected during rainy or humid conditions. Fungi invade seeds, especially in open storage, and reduce seed viability and vigor. Bujdoso (1979) recommends seed treatment with Ceresan, thiram, or captan to protect seeds from storage fungi and also to maintain high germination.

Invigoration of Stored Seeds

Seed germination and seedling vigor decrease with an increase in storage period. Use of certain chemical treatments during midstorage revives vigor. Solanki and Joshi (1985) reported that seed treatment with potassium orthophosphate (3 percent) improves the seed germination of four-year-old cucumber seeds. Seed coating with calcium chloride (40 millimolar [mM]) had a similar effect (Meng, Cheng, and Cui, 1996).

MUSKMELON: *Cucumis melo* L.

Introduction

Muskmelon is widely grown in the tropics. It is also known as sweet melon, cantaloupe, or honeydew melon. It is cultivated for its edible fruits, which have a sweet and musky flavor. Fruits are commonly used for dessert purposes. They are rich in sugar and vitamins A and C. Seeds can also be eaten after removing the hard seed coat.

Origin and Distribution

Muskmelon originated in tropical Africa. Subsequently, its cultivation spread to other tropical regions. It is largely grown in China, Egypt, France, India, Indonesia, Iran, Italy, Japan, Korea, Malaysia, Mexico, Morocco, the Philippines, Romania, Spain, Turkey, and the United States.

Morphology

Muskmelon belongs to family Cucurbitaceae and has diploid chromosome number 2n = 24. The plant is cultivated as an annual. It has long trailing vines, with leaves that are dark green, alternate, ovate, and three to seven lobed. Flowers are monoecious or andromonoecious. Pistillate flowers are borne singly, and staminate ones in clusters. Fruit is a fleshy pepo and varies in size, shape, and color. Fruits are globular or oblong, smooth or furrowed, green, yellow, or yellow-brown; flesh is green, pink, or yellow and has a musky scented flavor. Seeds are white or cream, black, or red-brown; flattened; and number 400 to 600 per fruit. One thousand seeds weigh about 5 to 25 g.

Seed Storage

Muskmelon seeds show orthodox storage behavior and withstand desiccation and chilling. Seeds remain viable for a long period under dry, cool conditions. High seed germination and vigor are retained for two years under ambient conditions. Seed storability varies in different cultivars (Doijode, 1987). Some cultivars are more sensitive to aging than others. Low-vigor seeds have higher respiratory quotients (Pesis and Ng, 1984).

Seed Collections

High-quality seeds with high viability and vigor are essential for high crop production. Crops grow well in sunny locations and during the warm season. Fruits produced under humid and rainy conditions are of inferior quality and get invaded by several diseases. Muskmelon requires deep, fertile, well-drained soil. High phosphorus content in the soil improves fruit quality and yield. Seeds are sown directly on ridges in rows or raised beds. Proper spacing, adequate nutrition, timely irrigation, and plant protection measures are followed. Muskmelon is compatible with other cucurbits and is generally pollinated by bees. To protect against genetic contamination, an isolation distance of 1,000 m is maintained. Off-type plants are removed, especially during the flowering and fruiting stages. Staminate flowers open early and outnumber pistillate flowers. Fruit color changes from green to yellow or white on maturity, and ripe ones are plucked for seed purposes. Seeds are separated from the pulp by washing and then are dried, cleaned, and packed for storage.

Fruit maturity. Seeds from fully matured fruits maintain high seed quality during storage. Fresh muskmelon seeds show dormancy, and an after-ripening period is required at all stages of seed development. Seeds are fully mature 60 days after anthesis. Seed germination increases in fully matured seeds after one year of storage at 20°C (Welbaum and Bradford, 1991). After-ripening improves seed quality and the yield of plants raised from the seeds (Ataullaev, 1982). During fruit ripening, the quantity of oligosaccharides and vitamin C content increase and monosaccharides, total acidity, and pectin content decrease. Only fruits 35 days old or older are suitable for after-ripening.

Fruit position. The position and number of fruits per plant influence seed quality. An increase in fruit number is accompanied by a decrease in fruit weight, number of seeds per fruit, seed weight, and germination percentage (Incalcaterra and Caruso, 1994).

Seed Germination

Seed germination is epigeal and completes in five days. Germination improves with pulp fermentation of 24 h. It was highest in fruits harvested at full maturity, as compared to those in the full-slip or half-slip stage (Singh, Lal, and Rana, 1985). Seeds from the proximal and middle portions of the fruit germinate better than seeds from the distal end. Seed germination and emergence hasten with storage up to 15 months and then decline, which is attributed to changes in lipase-1-peroxidase activity (Rakhimova and Rzhevskaya, 1973). Osmotic priming improves seedling emergence, whereas it has a deleterious effect on storage life of muskmelon seeds (Oluoch and Welbaum, 1996).

Storage Conditions

Seed moisture. Low seed moisture increases the storage life of seeds. Moisture content of 6.6 percent or less is ideal for long storage life of muskmelon seeds (Teotia, 1985). In ultradry storage, extreme drying damages muskmelon seeds, and at a higher level of seed moisture, the storability decreases, especially at room temperatures (Ji, Guo, and Ye, 1996).

Temperature. Seeds lose their viability rapidly under high temperatures, especially if they have high moisture content and are kept in open storage. Seeds maintain viability for two years at room temperatures, whereas the storability extends to eight years by storing them in polyethylene bags and laminated aluminum foil pouches at low (5°C) and subzero temperatures (−20°C) (Doijode, unpublished data).

Oxygen. A high level of oxygen reduces seed viability and vigor. Seeds stored in nitrogen atmosphere under ambient conditions exhibit higher germination percentage after six years of storage (Doijode, unpublished data).

Storage fungi. Seed quality is affected by fungal infection. Fungal activity is higher at high seed moisture and high humidity. Some of the fungi that associate with muskmelon seeds are *Botrytis cinerca, Cephalosporium* spp., *Cladosporium herbarium, Fusarium moniliforme, Mucor* spp., *Penicillium oxalicum,* and *Trichothecium* spp. Heat treatment at 45°C for 10 min eliminates several fungi, without any adverse effect on seed germination (Sharma and Roy, 1983). Also, fungicides such as captan and carbendazim are effective in controlling storage fungi.

WATERMELON: *Citrullus lanatus* (Thunb.) Mansf.

Introduction

Watermelon is a popular dessert fruit in the tropics and subtropics, especially in the dry regions of the world. It is grown for the edible juicy fruit. The sweet pulp, which is rich in sugars and amino acids, is eaten raw. The juice is canned and also used in the preparation of syrup. Seeds, which are rich in edible oil, are roasted with salt and eaten.

Origin and Distribution

Watermelon is native to south-central Africa. It is cultivated throughout the tropics and subtropics. It is grown in Brazil, China, Egypt, Greece, India, Iran, Iraq, Italy, Japan, Korea, Mexico, Morocco, Spain, Tunisia, Turkey, and the United States.

Morphology

Watermelon belongs to the Cucurbitaceae family and has diploid chromosome number 2n = 22. Watermelon is an annual dicotyledonous plant having vigorous spreading vines with a deep root system; thus, it can survive under relatively dry conditions. Leaves are pinnately lobed, number three to four pairs, and have hairy tendrils. Flowers are unisexual, single, axillary, light yellow, small, and less showy compared to other flowers of cucurbits. Male flowers exceed the number of female flowers. Fruit is large, oblong, globular, ellipsoidal, or spherical in shape, with a thick and fragile rind. Fruit color is green, or cream and striped. Flesh is juicy and pink to red, yellow, white or greenish white, and the central cavity is absent. Seeds are numerous, about 200 to 500 per fruit; brown, black, red, white, greenish, or yellow; and rich in carbohydrates, fats, and proteins. One thousand seeds weigh about 110 to 115 g.

Seed Storage

Watermelon is exclusively propagated by seeds. Seeds are also used in hybridization programs and in the conservation of genetic diversity. Seeds show orthodox storage behavior and are fairly good storers under ambient conditions. Seeds maintain their viability for a fairly long period (Nakamura, 1958). In a long-term study, seeds of several species of watermelon stored for 43 to 60 years at the National Seed Storage Laboratory, Fort Collins, Colorado, lost only 51 percent of their viability (Roos and Davidson, 1992).

Seed Collections

Watermelon is a warm-season crop. It requires a hot, dry climate, plenty of sunshine, and a long growing season. It grows well in deep, fertile, well-drained, sandy loam to loam soils. It is propagated by seeds. Seeds are sown directly in the field, either on hills or on ridges. Crops are suitably spaced, fertilized, watered, and protected against pests and diseases for better crop growth and seed production. Watermelon is a cross-pollinated crop, and pollination is mainly brought about by honeybees. Male flowers appear first, followed by pistillate flowers. An isolation distance of 1,000 m is provided for maintaining genetic purity. Off-type plants showing undesirable vegetative and fruit characteristics are removed, especially during the flowering and fruit development stages. Fruit takes about 80 to 110 days for maturity, and its color changes from green to pale yellow on the underside of the fruit. Seeds are located in the central areas of the fruit pulp. Seeds are removed along with pulp, washed, dried, cleaned, and processed for storage.

Fruit maturity. Fruit color changes on maturity and becomes light yellow, especially on the lower surface. During this stage, seeds also attain maturity and give higher germination. Watermelon fruit matures in 49 to 54 days. Normally, fruits are left on the vines for a comparable period after the vines become dry. In cucurbits, the seed coat completes its growth earlier than the embryo, and this may play an important role in inhibiting the germination of immature seeds. Seed coat removal increases germination in immature seeds (Nerson, 1991).

Fruit storage. Watermelon fruits can be stored for relatively long periods. Seeds from stored fruits also maintain seed quality for a long period. Bankole (1993) reported that high seed germination could be maintained for four months when fruits are stored under ambient conditions, but germination decreases on prolonged storage. Nerson and Paris (1988) observed that storage of mature fruits up to 48 months did not affect seed germination, but germination was affected in seeds of immature fruits.

Seed Germination

Watermelon seeds possess a hard seed coat and thus take longer to germinate. Seed germination is epigeal and slow at lower temperatures. It improves with simple washing of seeds with water and scarification of the seed coat. Kolev and Boyadzhiev (1983) reported that seed germination and storage life improved by treating the freshly extracted seeds with 2 percent hydrochloric acid or sodium hydroxide. Seed germinates readily on full imbibitions. Very dry or excess seed moisture affects the germination process. Seed germination is suppressed at minimum and maximum levels of 5 and 29 percent moisture (Earhart et al., 1979). Seed germination was higher under darkness, and intermittent exposure of seeds to red or far-red light affects the germination process (Loy and Evensen, 1979).

Storage Conditions

Seed moisture. High seed moisture is injurious to seed viability. Moisture content of 5 to 7 percent is safe for long-term storage. Excessive removal of moisture damages cell structure and affects seed quality. Ji, Guo, and Ye (1996) recommended moisture content of 1.25 to 4.26 percent for ultradry storage. Seed moisture less than 1.25 percent damages the cell structure, thereby reducing storability.

Seed packaging. A suitable container is required to maintain a desirable moisture level in seeds. Seeds can be stored in cloth bags, paper bags, glass containers, and laminated aluminum foil pouches, or aluminum cans, depending upon storage period and conditions. Polyethylene bags and laminated aluminum foil pouches are effective in maintaining high viability at lower temperatures (Doijode, 1994). Polyethylene bags are also cheaper, transparent, and sealable. Watermelon seeds stored in polyethylene bags at 2°C and 65 percent RH retain higher viability than seeds stored in paper bags (Kucherenko and Lebedeva, 1976). In a hermetically sealed box, seed viability was preserved for 15 to 23 years, and seeds packed in paper bags exhibited lower germination and more abnormal seedlings after four to five years of storage (Fursa and Zhukova, 1983).

Temperature. High temperature promotes seed deterioration (see Figure 42.2). Seeds stored at room temperatures remain viable for only two years, as compared to 15 years in those stored at 5 and –20°C (Doijode, unpublished data).

Air. Watermelon seed quality decreases during open storage under ambient conditions. A high level of oxygen promotes the respiration process. Seeds stored in nitrogen remain viable for seven years under ambient conditions (Doijode, unpublished data).

FIGURE 42.2. Influence of High Storage Temperatures and Relative Humidity on Seed Viability and Vigor in Watermelon

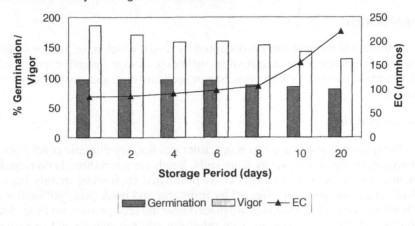

PUMPKIN:
Cucurbita moschata (Duchesne) Poir.

Introduction

Pumpkin is an important cucurbitaceous vegetable in arid regions of the tropics. It is cultivated for its edible fruits, which have a long storage life under ambient conditions. Ripe fruits are cooked as a vegetable and also are fed to cattle. Pumpkin is rich in carbohydrates, potassium, and vitamins A and C.

Origin and Distribution

Pumpkin originated in the arid regions of Central and South America. It is grown largely in the tropics. Some major pumpkin-growing countries are Argentina, China, Egypt, India, Indonesia, Iran, Italy, Japan, Malaysia, Mexico, Peru, the Philippines, South Africa, Spain, Turkey, and the United States.

Morphology

Pumpkin belongs to the Cucurbitaceae family. It is a dicotyledonous trailing plant having chromosome number 2n = 40. Leaves are simple, large, alternate, and deeply lobed. Plant is monoecious and bears a single yellow staminate flower on a long stem, while the pistillate is borne on a shorter stem. Fruit is large and varies in size, and its flesh color ranges from yellow to orange. Seeds are brown or white with a broad rim and contain about 30 percent proteins and 40 percent oils. One thousand seeds weigh about 200 g.

Seed Storage

Pumpkin is commercially propagated by seeds, which have greater value in genetic conservation. Seeds show orthodox storage behavior, and seed deterioration decreases with decreases in seed moisture and storage temperature.

Seed Collections

Pumpkin is a warm-season crop; it tolerates hot, dry climatic conditions. It requires deep, fertile, sandy loam soils. Seeds are sown directly on ridges or hills in the field. Pumpkin is cross-pollinated by insects, mainly bees. Fruit shape and yield are affected by improper and inadequate pollination. Therefore, one or two beehives are placed near the seed production field. An isolation distance of 1,600 m from other cucurbits is maintained to avoid cross-fertilization. Roguing of off-type plants is done during the flowering and fruiting stages. Only a few fruits per vine are retained for optimum size; otherwise, fruit weight will be reduced. Heavy fruits yield bold and high-quality seeds. Dematte (1982) reported that below average fruits give lower seed yield, but those seeds give the highest percentage of germination. Fruit takes normally four months from anthesis to maturity. Change of fruit color to yellow or straw color is the indication of maturity (see Figure 42.3).

FIGURE 42.3. Pumpkins Ready for Harvest

Seeds are removed along with the pulp from the matured fruit and subjected to fermentation in water for 24 h followed by washing in running water for 40 min (Zink and DeMendonca, 1965).

Fruit maturity. Seed viability and vigor are high in seeds harvested at the mature stage. Araujo, Mantovani, and Silva (1982) reported that pumpkin seeds extracted from fruits 15 to 65 days after anthesis can be stored for seven weeks. Seed germination and seedling vigor increase with increasing fruit age and storage length.

Fruit storage. Pumpkin fruits can be stored for a fairly long period at room temperatures. Seed quality improves with storage of fruits. Goldbach (1978) reported that seeds of ripe fruits give high viability and vigor. Seed germination in unripe fruits improves after ripening. Further, storage of ripe and overripe fruits at 23°C and 50 percent RH for six weeks resulted in loss of seed vigor and/or germination.

Seed Germination

Seed germination is epigeal, and seedling emergence takes less time at higher temperatures. Seed germination is better at alternate temperatures of 20/30°C for 16/8 h, respectively.

Seed size. Large seeds produce higher yield than small or medium-sized seeds, whereas there is no relationship between seed size at planting and seedling emergence (Wilson and Splittstoesser, 1979).

Chemicals. Seed germination in pumpkin improves by soaking seeds in an ammonium chloride solution (400 ppm) for 6 h (Singh and Singh, 1973).

Storage Conditions

Pumpkin seeds are viable for four years under ambient conditions, whereas high germination is retained for only two years (Doijode, 1995). Seed viability and vigor are affected by higher storage temperatures and relative humidity. Zink and DeMendonca (1965) preserved pumpkin seeds for 18 months by storing air-dried seeds in cloth bags under ambient conditions. Moisture content of 1.79 to 4.07 percent is safe for ultradry storage of pumpkin seeds. Further, reducing the moisture damages cell constituents, and higher moisture reduces seed storability (Ji, Guo, and Ye, 1996). Seed storage at 5 and –20°C maintains high viability and vigor for 15 years (Doijode, unpublished data). The loss of viability is associated with greater leaching of electrolytes, soluble sugars, and free amino acids and a decrease in dehydrogenase activity in seeds.

MARROW: *Cucurbita pepo* L.

Introduction

Marrow is a popular cucurbit in the tropics and subtropics. It is also known as summer squash or vegetable marrow. It is cultivated for its edible fruits. Immature fruits are boiled or roasted, and mature fruits are cooked as a vegetable. It is a good source of carbohydrates and vitamin A.

Origin and Distribution

Marrow originated in South America. Large genetic diversity exists in Mexico. It is cultivated in Argentina, Brazil, India, Indonesia, Malaysia, Mexico, the Philippines, and the United States.

Seed Storage

Marrow seeds exhibit orthodox storage behavior. These seeds can be dried to a very low moisture content, and storing them at chilling temperatures can extend longevity. Seeds are viable for two to three years under ambient conditions and need cool, dry conditions for better storage.

Morphology

Marrow belongs to family Cucurbitaceae and has chromosome number 2n = 40. Marrow is a monoecious annual that produces both bushy and long-running plants. Stems are rough with large, deeply lobed leaves. Flowers are bright yellow to orange yellow, single, and showy. Fruits differ in shape, size, and color and have a hard or soft shell. Fruits contain a large number of seeds that are easily separable from the pulp. Flesh is yellow or white. Seeds are brown, white, and large or small with a rim. One thousand seeds weigh 300 to 310 g.

Seed Collections

Marrow is a warm- and dry-season crop that withstands high temperatures. It grows well in well-drained, humus soil. It is propagated by seeds. Seeds are sown directly in the field, either on ridges or on hills. Plants are provided with adequate nutrients and watered regularly for better crop growth. Marrow is a cross-pollinated crop, and insects bring about pollination. Male flowers outnumber and appear earlier than female flowers. An isolation distance of 1,600 m is maintained to protect from cross-pollination with other cucurbits. Plants are inspected during the flowering and fruiting stages for various morphological characteristics, and off-type plants are removed. Fruit color changes on maturity. Dry fruits are collected, and seeds are removed from the pulp. Seeds are washed thoroughly, dried, cleaned, and packed for storage.

Fruit maturity. Seed germination is higher in seeds of matured fruits. Marrow seeds reach mass maturity 61 to 63 days after anthesis and moisture content decreases to 40 to 48 percent. Nerson and Paris (1988) reported that seeds extracted from immature fruits (28 days after anthesis) did not germinate. Seeds harvested 24 to 31 days after mass maturity gave higher seed longevity (Demir and Ellis, 1993).

Fruit storage. In cucurbits, fruit storage is a common practice used to improve seed quality. Storage of ripe fruits for three weeks and of immature fruits for 12 weeks at room temperature eliminates dormancy (Young, 1949). The mature marrow fruits can be stored for 48 months with germination unaffected (Nerson and Paris, 1988).

Seed Germination

Seed germination is epigeal and higher at alternate temperatures of 20/30°C for 16/8 h or at a constant temperature of 30°C (Solanki, Singh, and Yadav, 1980). Seeds take one week to germinate under dark conditions. Gomaa (1980) reported that seed placement at a 45° angle, with the embryo end upward, gives a higher percentage of germination and better seedling growth.

Seed dormancy. Marrow seeds exhibit dormancy to a certain extent. Seed dormancy is eliminated by after-ripening of seeds for one to two months (Ingold, 1960). Decoating of seeds also overcomes the dormancy and gives 100 percent germination shortly after mass maturity (Demir and Ellis, 1993).

Storage Conditions

Seed deterioration decreases at a lower level of seed moisture and at a lower temperature. Demir (1994) reported that seed longevity improves with a decrease in seed moisture and temperature. Seed moisture of about 6.5 percent gives highest seed storability, and any subsequent increase in moisture reduces seed longevity. Root growth is also adversely affected at higher seed moisture. Thus, marrow seeds are dried to a lower moisture content (5 to 7 percent) and suitably packed in moisture-proof containers, such as laminated aluminum foil pouches, and are preferably kept at –20°C for long-term storage. However, seeds can be packed in polyethylene bags and stored at 5°C for relatively longer storage.

Invigoration of Stored Seeds

Seed priming improves the rate of seedling emergence, produces uniform seedlings, and gives a higher germination percentage, irrespective of the osmotic agent used or the duration of treatment (Mauromicale, Cavallaro, and Ierna, 1994).

RIDGE GOURD: *Luffa acutangula* (L.) Roxb.

Introduction

Ridge gourd is also called angled gourd, angled loofah, Chinese okra, and silky gourd. It is cultivated in the tropics for its tender edible fruits. Fruits are cooked as a vegetable. Seeds are rich in protein and oil.

Origin and Distribution

Ridge gourd originated in India. Wide genetic variation for various morphological fruits is observed in different parts of India. It is cultivated in India, Indonesia, Malaysia, Myanmar, the Philippines, Sri Lanka, and Taiwan.

Morphology

Ridge gourd belongs to family Cucurbitaceae and has chromosome number 2n = 26. The plant is a dicotyledonous, monoecious, annual climber with tendrils. Leaves are pale green, simple, and ovate. Staminate flowers are many and arranged in a raceme, while pistillate flowers are solitary and fragrant. Sometimes hermaphrodite types also exist. Fruit is club shaped and angled, with ten prominent ribs, and many seeded. It becomes hard, bitter, and inedible on maturity. Seeds are black, flattened, and wrinkled and contain a high amount of oil. One thousand seeds weigh about 150 to 170 g.

Seed Storage

Ridge gourd seeds show orthodox storage behavior. Seed storage is enhanced by drying seeds and placing them under cooler conditions. Seeds are viable for two to three years under room temperatures. Higher storage temperature and relative humidity reduce storage life.

Seed Collections

Ridge gourd is a warm-season crop. It grows well in hot, humid climates. It prefers deep, humus, sandy loam soil. Seeds are sown directly in the field, either on hills or on ridges. Sufficient farmyard manure and chemical fertilizers are applied based on soil conditions. Timely irrigation is given, especially during the dry season at the flowering and fruit development stages. It is a cross-pollinated crop, and pollination is brought about by insects. An isolation distance of 800 m is provided to prevent genetic contamination. Plants are checked for various morphological characteristics during the flowering and fruiting stages, and off types are removed. Fruit matures in four to five months and changes color on maturity. Dry fruits are collected, and seeds are separated.

Seed Germination

Seed germination is epigeal and better at alternate temperatures of 20/30°C for 16/8 h, respectively. Seed germination is rather slow and uneven and takes longer to complete, due to the hard seed coat. Fursa and Gvozdeva (1971) noted that moist stratification of seeds at 30 to 35°C for three days and warm stratification at 40 to 50°C for two or three weeks gives higher percentage of germination.

Storage Conditions

High seed viability and vigor are maintained under proper storage conditions. These well-stored seeds germinate within five days, give better seedling growth and development, induce flowering early, and give a higher yield than low-vigor seeds (Witchwoot and Lavapaurya, 1985). Ridge gourd seeds were successfully preserved in aluminum foil pouches at 20 to 33°C and 67 to 86 percent RH for ten months, contrary to many reports claiming viability losses with packets (Villareal, Balagedan, and Castro, 1972). Further, well-dried seeds (5 to 7 percent mc) packed in polyethylene bags and stored at 5°C maintain high viability and vigor during storage. For long-term storage, it is preferable to pack low-moisture seeds in laminated aluminum pouches and to store them at subzero temperatures (–20°C).

SPONGE GOURD: *Luffa cylindrica* (L.) Roem

Introduction

Sponge gourd is also called dishcloth gourd, hechima, rag gourd, smooth loofah, and vegetable sponge. Sponge gourd is cultivated both for its use as a vegetable and for fiber purposes. The tender fruits of nonbitter types are cooked as a vegetable, and fiber is extracted from mature fruits. It is a good source of carbohydrates, calcium, phosphorus, iron, and vitamins A and C.

Origin and Distribution

Sponge gourd is native to India. Subsequently, it spread to Southeast Asian countries and other tropical regions. It is cultivated in Argentina, Brazil, India, Indonesia, Japan, Malaysia, the Philippines, and Sri Lanka.

Morphology

Sponge gourd belongs to family Cucurbitaceae and has diploid chromosome number $2n = 26$. It is a dicotyledonous, vigorous, annual climbing vine. Stems are rough and five angled. Leaves are dark green with five to seven lobes. Plant is monoecious, and male and female flowers are borne separately on the same vine. Male flowers appear in clusters and number from 4 to 20. Female flowers are single and axillary. Fruit is cylindrical with light stripes and many seeded. Seeds are black, flat, and smooth, with a narrow wing, and about 10 to15 mm long. They are rich in oil. One thousand seeds weigh about 90 to 100 g.

Seed Storage

Sponge gourd is primarily raised through seeds, which show orthodox storage behavior. Seeds are fairly good storers between growing seasons under ambient conditions.

Seed Collections

Sponge gourd requires a hot climate during growth. Excessive rainfall during flowering and fruiting affects yield as well as seed quality. It grows well in fertile, well-drained soil. Seeds are sown on ridges or on hills in the field. Plants are provided with adequate nutrients, timely irrigation, and plant protection measures. Plants are cross-pollinated by insects. A 1,000 m isolation distance is maintained between a sponge gourd field and other cucurbits. Fruits become yellow on maturity. Dry mature fruits are harvested, and the seeds are extracted, washed, dried, and suitably packed for storage.

Seed Germination

Seed germination is epigeal and higher at a constant temperature of 30°C or following scarification of seeds kept in moist sand at 30 to 35°C for three days (Fursa and Gvozdeva, 1971), whereas Singh and Mathur (1975) obtained higher seed germination through warm stratification (40 to 50°C) for two to three weeks. Higher temperature during storage causes progressive destruction of seeds. In sponge gourd, seedling growth is better with large seeds, and germination is unaffected by seed size but is maximum in black seeds as compared to other colored seeds (Mangal, Singh, and Pandita, 1979).

Storage Conditions

Dry, cool storage conditions are ideal for long storage of seeds. Well-matured seeds are dried to 6 to 7 percent moisture, packed in polyethylene bags, and stored at a low temperature (5°C). Seeds are packed in moisture-proof containers, such as laminated aluminum foil pouches, and stored at subzero temperature (–20°C) for long storage.

BITTER GOURD: *Momordica charantia* L.

Introduction

Bitter gourd is an important vegetable crop in Southeast Asia. It is mainly cultivated for its bitter unripe fruits. It is also known as alligator pear, balsam pear, bitter cucumber, bitter melon, ku gua, and foo gwa. Immature fruits are cooked as a vegetable. Leaves and fruits are used in medicine. It is valued for its high calcium, phosphorus, iron, and vitamin C contents. Dipping the fruits in salt water reduces the bitterness.

Origin and Distribution

Bitter gourd originated in India. Large genetic diversity is found in different states of India. It is a popular vegetable crop in Brazil, China, India, Indonesia, Malaysia, the Philippines, Singapore, and Sri Lanka.

Morphology

Bitter gourd belongs to the Cucurbitaceae family and has chromosome number 2n = 22. It is an annual, monoecious, dicotyledonous climbing vine. Stems are slender with palmate, deeply-lobed, simple leaves. Flowers are small, yellow, and borne singly at the axil. Fruit is green, pear-shaped, with a smooth or pointed protrusion on the surface, and many seeded. It bursts on maturity, showing red arils. Seeds are brown, 1 to 1.5 cm long, and oval with flattened arils. One thousand seeds weigh 60 to 170 grams.

Seed Storage

Seeds are used in propagation and in crop improvement through hybridization. Seeds are a valuable tool for long-term conservation of germplasm. Bitter gourd seeds show orthodox storage behavior; they can tolerate desiccation and preserve their viability at lower temperatures.

Seed Collections

Bitter gourd grows well in hot, humid areas. It prefers fertile, well-drained soil that is rich in organic matter. It is propagated by seeds and sown directly in the field on ridges or hills. Plants are fertilized and protected against pests and diseases. Irrigation is provided during the dry season. It is a cross-pollinated crop, and different insects bring about pollination. Cross-pollination increases the percentage of large seeds in the fruit (Catedral and Mamicpic, 1976). An isolation distance of 1,600 m is provided to maintain genetic purity. Plants are checked for various morphological characteristics, especially during the flowering and fruiting stages, and abnormal ones are discarded. Fruit turns yellow or orange on ripening, and then it splits, exposing seeds. Ripe fruits are collected, and seeds are removed, washed thoroughly, dried, cleaned, and processed for storage.

Fruit maturity. Seeds continue to mature in the fruit, even if the fruit is harvested before ripening and stored under ambient conditions. Harvesting of fruits three days before full ripening and storing them under ambient conditions for four days gave high germination percentage and vigor; this also avoided field losses due to bird or rain damage (Krishnaswamy, 1991).

Fruit size. Bitter gourd fruit size varies in different cultivars and depends on growing conditions. Seeds from very long fruits (30 cm) give higher seed germination and fruit quality, whereas germination is poor in seeds from small fruits (15 cm) (Vanangamudi and Palaniswamy, 1989).

Seed Germination

Seed germination is epigeal and higher at a constant temperature of 30°C or at alternate temperatures of 20/30°C for 16/8 h. Seed germination is rapid and greater in water-soaked seeds. Boron application or rupturing the hard seed coat can improve germination (Singh, Vashistha, and Singh, 1973).

Seed dormancy. Bitter gourd seeds show seed coat dormancy and endogenous dormancy. Fresh seeds do not germinate readily, and they require after-ripening for one month (Devi and Selvaraj, 1994). Application of potassium nitrate (1 percent) eliminates the dormancy and promotes seed germination.

Light. Light inhibits the germination process in bitter gourd (Nakamura, Okasako, and Yamado, 1955). However, daylight along with gibberellic acid (50 ppm) application enhances seed germination. Further, seed exposure to red light improves the germination, unlike exposure to green and yellow light, which showed an inhibitory effect (Nath, Soni, and Charan, 1972).

Growth substances. Seeds soaked in a low concentration of gibberellic acid (100 to 200 ppm) and indoleacetic acid (100 to 200 ppm) showed improved seed germination and seedling emergence (Sharma and Govil, 1985).

Seed position. Position of seeds placed in soil affects emergence of seedlings. Seeds placed vertically give better seedling emergence and show higher seedling vigor (Krishnaswamy, 1992).

Storage Conditions

Bitter gourd seeds remain viable for two to three years at room temperatures. Seeds with 5 to 7 percent moisture maintain high viability at low temperatures. Seeds can be stored at 5°C in polyethylene bags, but for long storage, seeds are packed in laminated aluminum foil pouches and stored at −20°C. Seed deterioration is minimal under cool, dry storage conditions. The loss of seed vigor precedes a decrease in viability and is manifested in a greater leaching of metabolites from seeds. Seed germination was reduced to 62 percent on aging of seeds, and this was positively correlated with the leaching of soluble sugars and amino acids (see Table 42.1) (Doijode, 1990).

BOTTLE GOURD:
Lagenaria siceraria (Mol.) Standl.

Introduction

Bottle gourd is an important gourd having a wide range of uses. It is largely cultivated in the tropics and subtropics for its edible fruits. It is also known as calabash gourd, trumpet gourd, white-flowered gourd, and zucca melon. It is a good source of carbohydrates and calcium. Tender fruits are cooked as a vegetable, and the hard shell is used in making various types of

TABLE 42.1. Effect of High Temperature (42°C) and High Humidity (95 percent) on Seed Quality in Bitter Gourd

Storage Period (days)	Germination (%)	Vigor Index	Electrical Conductivity (mmhos)	Soluble Sugars (mg·g⁻¹)	Free Amino Acids (mg·g⁻¹)
0	90	1527	455	1.93	8.41
2	82	1382	466	2.04	8.90
4	72	1206	481	2.08	9.18
6	66	1097	532	2.25	10.18
8	62	1007	546	2.52	10.81

utensils for storage of liquid and food materials. It is also used in designing certain musical instruments. The pulp and leaves have medicinal properties.

Origin and Distribution

Bottle gourd originated in south-central Africa and probably later spread to South America, where it exhibits a certain amount of genetic variability. It is cultivated in Brazil, China, Colombia, Egypt, France, India, Indonesia, Japan, Malaysia, the Philippines, Spain, Sri Lanka, and the United States.

Morphology

Bottle gourd belongs to family Cucurbitaceae and has chromosome number 2n = 22. It is a monoecious plant with long climbing or trailing vines. Leaves are simple, large, pale green, and hairy. Flowers are single, axillary, and white; borne on a long peduncle; fragrant; and short-lived. Fruit varies in size and shape, being flat, globular, or bottle or club shaped, and green or white in color. Rind becomes hard on maturity. Seeds are dicotyledonous, obovate, about 2 cm long, white or brown, and ridged. One thousand seeds weigh about 130 to 150 g.

Seed Storage

Bottle gourd is propagated commercially through seeds. Farmers preserve seeds in viable condition between growing seasons, whereas breeders want to store them for very long periods without affecting the genetic architecture of the genotypes. Both types of storage demand proper care and suitable storage conditions. Bottle gourd seeds lose viability under ambient conditions. Viability decreases to 50 percent after 12 months, and none of the seeds germinated after 24 months of storage under ambient conditions (Doijode, 1989).

Seed Collections

Preharvest factors play an important role in seed storage. Seeds obtained from healthy, normal crops are vigorous and store for comparatively long periods. Bottle gourd gives optimum yield with moderate rainfall, sunny days, and warm climate. It grows well in sandy loam soils rich in organic matter. Seeds are sown directly in the field on ridges or mounds. Optimum fertilizer and regular irrigation during the dry period produce healthy crop growth. Bottle gourd is cross-pollinated by insects, and an isolation distance of 800 m is maintained. Off-type plants are removed to safeguard genetic purity. Mature and dry fruits are harvested, and the seeds are removed, cleaned, dried, and stored.

Seed Germination

Fresh bottle gourd seeds show a certain amount of dormancy. It takes two to three months of after-ripening to eliminate the dormancy (Nakamura, Yamada, and Shimizu, 1978). Seed scarification and/or sowing at alternate temperatures of 20/30°C for 16/8 h enhances germination. Sandy soil and a constant temperature of 30°C is ideal for optimum seed germination (Singh, Singh, and Khanna, 1973), and germination is nil at 35°C (Solanki and Seth, 1981). Seeds placed vertically with the embryo end facing upward give early and higher seedling emergence (Krishnaswamy, 1992). Seed germination improves with soaking seeds in gibberellic acid at 200 ppm (Pampapathy, Sriharibabu, and Narasimharao, 1989) and in succinic acid (600 ppm) (Singh and Singh, 1976). Nath and Mathur (1967) observed that seeds exposed to red light gave higher germination than those exposed to blue or brown light.

Storage Conditions

Storing dry seeds in a cooler environment extends their longevity. Seeds packed in laminated aluminum foil pouches and stored at 20 to 33°C and 67 to 86 percent RH can be stored for ten months (Villareal, Balagedan, and Castro, 1972). Further, seeds remain viable for five years at 5 and −20°C. Seeds retain high viability when stored in polyethylene pouches at 5°C and in laminated aluminum foil pouches at −20°C (see Figure 42.4) (Doijode, 1998). Seeds retain high germination in polyethylene bags and laminated aluminum foil pouches for ten years (Doijode, unpublished data). On deterioration, seed germination and seedling vigor decrease. The soluble sugar and free amino acid contents are higher in leachates of aged seeds (Doijode, 1989).

FIGURE 42.4. Influence of Temperature on Seed Longevity in Bottle Gourd

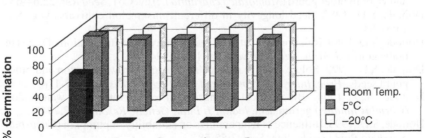

REFERENCES

Ali, N., Skirvin, R., Splittstoesser, W.E., and George, W.L. 1991. Germination and regeneration of plants from old cucumber seeds. *Hort Sci.* 26:917-918.

Araujo, E.F., Mantovani, E.C., and Silva, R.F.D. 1982. Influence of fruit age and storage period on squash seed quality. *Revista Brasileira de Sementes* 4:77-87.

Ataullaev, N.A. 1982. Seed quality and the characteristics of metabolism in melons in relation to fruit age. *Sel'skhozyaistvennaya Biol.* 17:509-512.

Bankole, S.A. 1993. Moisture content, mould invasion and seed germinability of stored melon. *Mycopathologia* 122:123-126.

Boos, G.V. 1966. Increasing seed viability in cucumbers and tomatoes under glasshouse conditions. *Trudy Priklad Bot. Genet. Selek.* 38:178-190.

Bujdoso, G. 1979. Effect of prestorage treatment on cucumber seed germination. *Zoldsegtermesztesi Kutato Intezet Bull.* 13:49-54.

Buriev, Kh.Ch. 1984. Certain germination characteristics of cucumber seeds of different maturity. *Nauchnye Trudy Tashkentskii Selskokhozyais tvennyi Institut* No. 105:70-76.

Cantliffe, D.J. and Phatak, S.C. 1975. Plant population studies with pickling cucumber grown for once over harvest. *J. Am. Soc. Hort. Sci.* 100:464-466.

Catedral, I.G. and Mamicpic, N.G. 1976. The effect of spacing and artificial pollination on seed yield and other characters of ampakya (*Momordica charantia* L.). *Philippine J. Crop Sci.* 1:189-190.

Cooper, C.C. 1959. Polyethylene protective seed packages. *Proc. Am. Soc. Hort. Sci.* 74:569-579.

Cui, H.W. and Yin, Y.G. 1995. Effect of hot treatment on the vigor of newly harvested cucumber seeds. *Rep. Cucurbit Genet. Coop.* No. 18:17-18.

Dematte, M.E.S.P. 1982. Yield components of mature fruits and seeds of *Cucurbita moschata* cv. Canhao IAC 3046. *Proc. Trop. Reg. Am. Soc. Hort. Sci.* 25:47-50

Demir, I. 1994. Effect of seed moisture content on storage longevity of marrow (*Cucurbita pepo* L.). *Bahce* 23:53-58.

Demir, I. and Ellis, R.H. 1993. Changes in potential seed longevity and seedling growth during seed development and maturation in marrow. *Seed Sci. Res.* 3:247-257.

Devi, J.R. and Selvaraj, J.A. 1994. Effect of presowing treatment on germination and vigor in bitter gourd (*Momordica charantia* L.) cv. Co1. *Seed Res.* 22:64-65.

Doijode, S.D. 1987. Seed longevity in different muskmelon cultivars. *Veg. Sci.* 14:51-54.

Doijode, S.D. 1989. Relationship between seed quality and deterioration of seeds in bottle gourd. *J. Mah. Agril. Universities* 14:242-243.

Doijode, S.D. 1990. Solute leakage on accelerated aging in watermelon and bitter gourd seeds. *Proc. Inter. Sat. Symp. Seed Sci. Technol. Hisar* 1:107-110.

Doijode, S.D. 1994. Biochemical changes and conservation of watermelon (*Citrullus lanatus*) germplasm. *Indian J. Plant Genetic Resources* 7:79-84.

Doijode, S.D. 1995. Influence of temperatures on storability of pumpkin (*Cucurbita moschata* Poir.) seeds. *Veg. Sci.* 22:55-58.

Doijode, S.D. 1998. Conservation of bottle gourd (*Lagenaria siceraria* Standl.) germplasm through seed storage. *Indian J. Plant Genetic Resources* 11:56-58.

Earhart, D.R., Fuqua, M.C., Tereskovich, G., and Downes, J. 1979. The effect of temperature and moisture levels on germination of the triploid watermelon. *HortSci.* 14:123.

Edwards, M.D., Lower, R.L., and Staub, J.E. 1986. Influence of seed harvesting and handling procedures on germination of cucumber seeds. *J. Am. Soc. Hort. Sci.* 111:507-512.

Eguchi, T. and Yamada, H. 1958. Studies on the effect of maturity on longevity in vegetable seeds. *Natl. Inst. Agr. Sci. Bull. Ser. E. Hort.* 7:145-165.

Fedosenko, V.A. and Yuldasheva, L.M. 1976. The preservation of *Cucumis sativus* seeds at extremely low temperatures. *Bull. Vsesoyuznogo Ordena Lenina Instituta Rastenievodstva imeni* No. 64:60-62.

Frohlich, H. and Henkel, A. 1964. The problems of the duration of viability and the quality of outdoor cucumber seeds. *Die Gartenbauwissenschaft* 11:130-132.

Fursa, T.B. and Gvozdeva, Z.V. 1971. Increasing of seed germination rate in some species of the family Cucurbitaceae. *Trudy po Prikladnol Botanike Genetike i selektsii* 44:211-214.

Fursa, T.B. and Zhukova, N.V. 1983. Viability of watermelon seeds under conditions of natural and accelerated aging. *Trudy po Prikladnol Botanike Genetike i selektsii* 81:52-57.

Goldbach, H. 1978. Vigor of *Cucurbita moschata* seeds in relation to ripening stage as measured by the accelerated aging tests. *Turrialba* 28:343-345.

Gomaa, H.M. 1980. Influence of seed orientation in seedbed on emergence and seedling growth of some cucurbits. *Egypt. J. Hort.* 7:55-62.

Incalcaterra, G. and Caruso, P. 1994. Seed quality of winter melon *(Cucumis melo var. indorus)* as influenced by the position of fruits on the mother plant. *Acta Hort.* 362:113-116.

Ingold, M. 1960. Contribution a lethde de la germination des Semences d' *Allium cepa* L. et *Cucurbita pepo* L. *Proc. Inter. Seed Test. Assoc.* 25:787-799.

Ji, Z.X., Guo, C.G., and Ye, Y.F. 1996. Study on ultra dry storage in four melon seeds. *Acta. Agril. Zhejiangensis* 8:50-53.

Kolev, E. and Boyadzhiev, K.H. 1983. Possibilities of washing seeds of fleshy fruited vegetables with chemicals. *Gardinarska i Lozavska Nauka* 20:44-49.

Kononkov, P.F., Kravchuk, V.Ya., and Vasyanova, A.V. 1975. The effect of high temperature and relative humidity on the changes in seed quality of vegetable crops. *Dokl. Vseso. Ord. Len Akad. Sel. Nauk. imeni Lenina* No. 7:15-16.

Krishnaswamy, V. 1991. Post-harvest seed maturation in bitter gourd. *J. Appl. Seed Prod.* 9:41-43.

Krishnaswamy, V. 1992. Effect of orientation of seed placement in soil on seedling emergence in some cucurbitaceous vegetables. *Seed Res.* 20:70-73.

Kucherenko, N.E. and Lebedeva, N.N. 1976. The effect of the storage regime and the packing method on the germination of cucurbit seeds. *Vopro Trop. Sub-Tropic. S-Kh. Moscow*, pp. 47-51.

Kurdina, V.N. 1966. Changes in the sowing quality of seeds of vegetable crops during storage. *Izv. timirjazev sel-hoz. Akad.* No. 5:135-144.

Loy, J.B. and Evensen, K.B. 1979. Phytochrome regulation of seed germination in a dwarf strain of watermelon. *J. Am. Soc. Hort. Sci.* 104:496-499.

MacNeill, M.M. and Hall, B. 1982. Factors affecting the germination of cucumber seeds *(Cucumis sativus* L.). *Plantsman* 3:251-253.

Mangal, J.L., Singh, K.P., and Pandita, M.L. 1979. Effect of seed size and seed coat color on seed quality in sponge gourd. *J. Res.* 9:14-18.

Mauromicale, G., Cavallaro, V., and Ierna, A. 1994. Effects of seed osmoconditioning on emergence characteristics of the summer squash *(Cucurbita pepo* L.). *Acta Hort.* 362:221-228.

Meng, H.W., Cheng, Z.H., and Cui, H.W. 1996. Effects of calcium on cucumber *(Cucumis sativus)* seed germination, seed storage and seedling growth. *Rep. Cucurbit Genetics Coop.* No. 19:23-24.

Nakamura, H., Yamada, H., and Shimizu, T. 1978. Several factors affecting the heat resistance of the seeds of bottle gourd *(Lagenaria siceraria). Bull. Veg. Orn. Crop Res. St. Japan* 4:119-147.

Nakamura, S. 1958. Storage of vegetable seeds. *J. Hort. Assoc. Japan* 27:32-44.

Nakamura, S., Okasako, Y., and Yamado, Y. 1955. Effect of light on the germination of vegetable seeds. *J. Hort. Assoc. Japan* 24:17-28.

Nandeesh, Javaregowda, S., and Ramegowda. 1995. Studies on the storage of harvest and postharvest ripening on seed quality in cucumber *(Cucumis sativus). Seed Res.* 23:113-115.

Nath, P. and Mathur, N.K. 1967. Effect of plant growth regulators, light treatment and their interactions on seed germination in bottle gourd *Lagenaria siceraria. Proc. Inter. Symp. Sub-Trop. Trop. Hort. Bangalore* 1:396-403.

Nath, P., Soni, S.L., and Charan, R. 1972. Effect of plant growth regulators, light and their interactions on seed germination in bitter gourd *(Momordica charantia). Proc. 3rd Inter. Symp. Sub-Trop. Trop. Hort.* 2:173-182.

Nerson, H. 1991. Fruit age and seed extraction procedures affect germinability of cucurbit seeds. *Seed Sci. Technol.* 19:185-195.

Nerson, H. and Paris, H.S. 1988. Effects of fruit age, fermentation and storage on germination of cucurbit seeds. *Scientia Hort.* 35:15-26.

Odland, M.J. 1937. Observations on dormancy in vegetable seeds. *Proc. Am. Soc. Hort. Sci.* 35:562-566.

Oluoch, M.O. and Welbaum, G.E. 1996. Viability and vigor of osmotically primed muskmelon seeds after nine years of storage. *J. Am. Soc. Hort. Sci.* 121:408-413.

Pampapathy, K., Sriharibabu, R., and Narasimharao, C.L. 1989. Studies on the effect of seed treatment with GA_3 on germination of cucurbitaceous vegetables. *Andhra Agric. J.* 36:294-297.

Pesis, E and Ng, T.J. 1984. The role of anaerobic respiration in germinating muskmelon seeds. I. In relation to seed lot quality. *J. Exp. Bot.* 35:356-365.

Rakhimova, R.S. and Rzhevskaya, F.Yu. 1973. The quality of melon seeds in relation to the time of fruit ripening, position on the placenta and duration of storage. *Ovoshche Bakh. Kultur Kartofelya* No. 10:15-26.

Rodrigo, P.A. 1953. Some studies on the storing of tropical and temperate seeds in the Philippines. *13th Inter. Hort. Congr.* Rpt. 2:1061-1066.

Roos, E.E. and Davidson, D.A. 1992. Record longevities of vegetable seeds in storage. *HortSci.* 27:393-396.

Sharma, A.K. and Govil, C.M. 1985. Response to growth substances of seed germination, seedling growth and hypocotyl anatomy in *Momordica charantia* L. *Veg. Sci.* 12:1-6.

Sharma, R.K. and Roy, A.N. 1983. Physiotherapeutic treatments of muskmelon seeds. *Seed Res.* 11:187-190.

Shifriss, O. and George, W.L. 1965. Delayed germination and flowering in cucumber. *Nature* 206:424-425.

Singh, A. and Singh, H.N. 1973. Note on the effect of presoaking seeds in N solution on germination and early seedling growth in pumpkin *(Cucurbita moschata)*. *Indian J. Agric. Sci.* 43:973-976.

Singh, A. and Singh, H.N. 1976. A note on effect of presoaking of seeds in nitrogen and succinic acid solution on germination and seedling growth in bottle gourd. *Prog. Hort.* 7:35-38.

Singh, B., Vashistha, R.N., and Singh, R.N. 1973. Note on the effect of certain chemicals on seed germination of bottle gourd, bitter gourd, watermelon and bhindi. *Haryana J. Hort. Sci.* 2:70-71.

Singh, H., Lal, T., and Rana, R.S. 1985. Effect of fruit maturity and fermentation on seed germination of muskmelon *(Cucumis melo)*. *Seed Res.* 13:171-175.

Singh, M.M. and Mathur, M.K. 1975. Effect of temperature and moisture stresses on germination of seeds. IV. Studies on *Luffa cylindrica*. *Veg. Sci.* 2:87-92.

Singh, P.V., Singh, M.B., and Khanna, A.N. 1973. Germination studies on bottle gourd *(Lagenaria siceraria)*. *Seed Res.* 1:63-66.

Solanki, S.S. and Joshi, R.P. 1985. Effect of different chemicals on invigoration in seed germination of cucumber *(Cucumis sativus)* and capsicum *(Capsicum annuum)*. *Prog. Hort.* 17:122-124.

Solanki, S.S. and Seth, J.N. 1981. A note on the physical factors affecting coefficient velocity of germination of bottle gourd and methi seeds. *Prog. Hort.* 13:57-60.

Solanki, S.S. and Seth, J.N. 1984. Studies on coefficient velocity of germination of winged bean *(Psophocarpus tetragonolobus)* and cucumber *(Cucumis sativus* L.*)* seeds. *Haryana J. Hort. Sci.* 13:143-146.

Solanki, S.S., Singh, R.D., and Yadav, J.P. 1980. Studies on the temperature and media relations and coefficient velocity of germination of vegetable seeds. II. Summer squash *(Cucurbita pepo)* and okra *(Abelmoschus esculentus)*. *Prog. Hort.* 12:59-65.

Staub, J.E., Wehner, T.C., and Tolla, G.E. 1989. The effect of chemical seed treatment on horticultural characteristics in cucumber *(Cucumis sativus)*. *Scientia Hort.* 38:1-10.

Sunil, G.D.J.L. and Mabesa, R. 1991. Storage techniques for extending the viability of pregerminated cucumber seeds. *Trop. Agril.* 147:19-31.

Teotia, M. S. 1985. Sorption behavior of melon seed kernels. *J. Food Sci. Technol.* 22:283-285.

Vanangamudi, K. and Palaniswamy, V. 1989. Effect of fruit grading on seed quality characteristics in bitter gourd. *Veg. Sci.* 16:96-98.

Villareal, R.L., Balagedan, J.B., and Castro, A.D. 1972. The effects of packing materials and storage conditions on the vigor and viability of squash *(Cucurbita maxima)*, patola *(Luffa acutangula)*, and upo *(Lagenaria siceraria)* seeds. *Philippine Agriculturist* 56:59-76.

Wallerstein, I.S., Goldberg, Z., and Gold, B. 1981. The effect of age and fruit maturation on cucumber seed quality. *Hassadah* 61:570-574.

Wehner, T.C. 1982. Genetic variation for low temperature germination ability in cucumber. *Rep. Cucurbit Genetics Coop.* No. 5:16-17.

Welbaum, G. and Bradford, K.J. 1991. Water relations of seed development and germination in muskmelon *(Cucumis melo)*. VII. Influence of after-ripening and aging on germination response to temperature and water potential. *J. Exp. Bot.* 42:1137-1145.

Wilson, M.A. and Splittstoesser, W.E. 1979. Effect of pumpkin seed size on seedling emergence and yield on two soil types. *HortSci.* 14:731.

Witchwoot, S. and Lavapaurya, T. 1985. Seed vigor tests and effect of seed vigor on growth and yield of the ridge gourd (*Luffa acutangula* Roxb.). *Proc. 23rd Nat. Conf. Kasetsart University* 1:163-177.

Yin, M.G. and Cui, H.W. 1995. Alteration of catalase activity and ethylene release during germination in newly harvested cucumber seeds. *Rep. Cucurbit Genet. Coop.* No. 18:19-20.

Young, R.E. 1949. The effect of maturity and storage on germination of butternut squash seed. *Proc. Am. Soc. Hort. Sci.* 53:343-346.

Zaitseva, G.A. 1972. The effect of conditions and duration of cucumber seed storage on crop quality. *Ovochvnitstvo i Bashtanitstvo Resp Mizhvid Temat Nauk* No. 14:46-49.

Zink, E. and DeMendonca, N.T. 1965. The effect of the extraction method on the viability of pumpkin seeds. *Bragantia* 24:1-2.

SECTION IV:
SEED STORAGE
IN ORNAMENTAL CROPS

Ornamental Crops

Many ornamental crops are propagated commercially by vegetative means. However, seeds are also used to a limited extent. Research on the storage of flower seeds is scarce predominantly because it is undertaken by flower seed companies and is not published. Grzesik (1993) reported that lowering storage temperature and relative humidity extended seed longevity considerably in several ornamental species. Storage of low-moisture seeds (4 to 7 percent) at temperatures below 0°C in sealed containers is ideal for longer storage (Bass, 1980). Carpenter, Ostmark, and Cornell (1995) preserved several ornamental seeds at 5, 15, and 25°C and 11 to 75 percent RH. Seed germination decreases with an increase in seed moisture content during storage. Likewise, Benedetto and Tognetti (1985) observed retention of high germination in many flower seeds stored under diffuse illumination or in darkness.

ROSE (*Rosa* spp.)

Introduction

The rose is an important commercial flower, widely appreciated and valued for its beauty, color, and fragrance. Roses are largely used as cut flowers and planted in beds and gardens; rose shrubs are also used in landscaping. The present-day cultivars are derived from the natural hybridization of Chinese and European roses. Roses are broadly grouped into hybrid tea, floribunda, polyantha, climber and rambler, miniature, and shrub roses. Some of the genotypes bear attractive colored fruits called hips. Flowers of hybrid tea and floribunda roses are largely used as cut flowers.

The rose is a perennial, spiny, deciduous plant. Leaves are alternate and pinnate. Flowers are solitary or borne in clusters, and they vary in size and color, namely, white, yellow, orange, pink, red, and lavender. Fruit is a fleshy hip that contains hairy achenes. Rose grows well in temperate conditions and is propagated by seeds, cuttings, budding, and grafting. Seeds are normally used for evolving new cultivars through hybridization, while budding is predominantly employed for propagation.

Seed Storage

Seed Germination

Preservation of high seed viability and seedling vigor is the main criterion for seed storage. In nature, rose plants set seeds poorly and in small numbers. Rose seeds do not germinate immediately after harvest due to the impervious nature of their seed coat and the presence of innate dormancy. A certain period of after-ripening is necessary before the seeds are ready to germinate. Seeds are extracted from mature fruits when their color changes from green to red for higher germination. Crushing the hips and removing the fleshy parts improves seed germination (Haenchen, 1968). Seeds from older hips germinate better than those from younger ones (Foster and Wright, 1983). Hips about 15 to 20 weeks old show a higher germination percentage (Mckelvie and Walker, 1975). Seed scarification, temperature stratification, and applications of growth substances hasten the germination process. Moist seed storage at 4°C for ten weeks overcomes the cold requirements. Seed soaking in a zinc and hydrochloric acid mixture softens the pericarp and improves the germination percentage. Seed coat removal hastened germination by seven days at 20°C, but it is a cumbersome process (Mckelvie and Walker, 1975). The combination of warm and cold stratification in a sequence improves the germination process in certain *Rosa* species. Seeds of *R. canina* preserved at 26.7°C for eight weeks followed by storage at 4.4°C for 12 weeks resulted in higher germination, while seeds of *R. laevigata* did not respond to warm or cold stratification and required a long storage period for germination. Popcov and Buc (1966) obtained high germination by soaking seeds in concentrated sulfuric acid for 1 h followed by warm stratification (20 to 25°C for two to three months) and cold stratification (5 to 6°C for three to five months). Blundell and Jackson (1971) suggested soaking the seeds in concentrated sulfuric acid for 1 h and then storing them at 26°C for three months followed by storage at 2°C. Yu, Yeam, and Kim (1975) reported that cold stratification for 60 days is sufficient for higher germination of *R. multiflora* seeds. Seed treatment with growth substances promotes the germination process. Seed soaking in gibberellic acid (10,000 mg·liter^{-1}) and benzyladenine (100 mg·liter^{-1}), followed by warm (20°C) and cold (5°C) stratification for four weeks, yielded higher germination (Foster and Wright, 1983). Presowing seed treatment with vitamins B_1, B_2, and C increases the germination as well as seedling vigor in seeds of *R. canina* (Serebryakova and Kalanova, 1977).

Storage Conditions

Rose seeds preserve well under ambient conditions. Seed deterioration is slow in different *Rosa* species (Crocker and Barton, 1931). The presence of seed dormancy is beneficial for preserving viability during storage. Rose seeds exhibit orthodox storage behavior. Seed longevity improves with a reduction in moisture and storage at low temperature. Seeds of certain *Rosa* species were preserved in dry conditions for two to four years. Seeds of *R. rugosa* had better germination after three years of storage. The percentage of germination dropped in seeds of *R. multiflora* from 72.5 to 48.4 in three years and in seeds of *R. setigera* from 53.4 to 35.6 after two years of storage (Crocker and Barton, 1931). Harrington (1968) indicated that a rise in seed moisture and in storage temperature reduces the storage life of seeds. Before storage, seeds are thoroughly washed in water, and all adhering pulp material is removed. Later, seeds dried to 5 to 6 percent moisture are preferably stored at low temperature (15°C) and low humidity (10 percent RH) for safer and longer storage (Cromarty, Ellis, and Roberts, 1982). Dried seeds packed in suitable moisture-proof containers, such as thick polyethylene bags and aluminum foil or cans, and stored at a low temperature can preserve high seed viability and vigor during storage.

ORCHID

Introduction

Orchids are the most beautiful flowers in the biosphere because of their unique shape and attractive coloring. They vary in shape, size, and color. These flowers bloom on the plants over a period of about three to four weeks. Orchids are grown in gardens and in pots and are used as cut flowers. Many orchids of different genera and species flourish well in humid, tropical conditions. These are grown predominantly in Australia, India, Malaysia, Mexico, Myanmar, New Guinea, the Philippines, Sri Lanka, Thailand, and the United States. Orchids are either terrestrial or epiphytic, the latter being in the majority and widely distributed in forested areas. The important cultivated orchid genera are *Cymbidium, Dendrobium, Paphiopedilum, Phalaenopsis, Pleione,* and *Vanda.*

Orchid is a monocotyledonous, herbaceous, perennial plant. Flowers vary in shape, size, and color, and they are fragrant in certain species. Orchid is propagated both sexually and asexually. The former causes variation in various plant characteristics and hence is used sparingly in commercial propagation. Cuttings, offsets, or meristem cultures are used for orchid propagation. Seeds are used in evolving new genotypes through hybridization and also in the conservation of genetic diversity in the seed bank. Orchid seeds demand aseptic conditions while raising seedlings, and in nature, germination requires the help of fungi. Each capsule contains about two to three million minute seeds, and only a small percentage (0.2 to 0.3 percent) germinate.

Seed Storage

Seed Germination

Seed germination is very low and slow in orchids. Seeds germinate in association with mycorrhizal fungi, mostly nonsporal types, such as *Rhizoctonia languinosa, R. mucoroides,* and *R. repens.* Orchid seeds lack endosperm, and they depend on other sources such as fungi for food. Orchid seeds are dependent on extraneous source for their sugar supply, particularly during germination (Smith, 1974). Seed germination is higher and faster in a sugar medium, especially the Knudson's C solution. Linden (1980) reported that immature seeds of orchid germinate better and show higher seedling growth than dried mature seeds in two mineral media: Fasts' and Burgeff-Egl. Seedlings grow better with powdered bark of incense cedar *(Libocedrus decurrens)* or sugar pine *(Pinus lambertiana)* added to the medium (Frei, Fodor, and Haynick, 1975). Mature seeds germinate earlier and develop protocorms (Hossain, Mannan, and Roy, 1996). The germination percentage was higher (75 percent) in 11-month-old seeds of *Vanda roxburghi.* Embryos of terrestrial orchid seeds are cultured in a medium containing macro- and microelements, sucrose, amino acids, and vitamins and are incubated in the dark at 25°C. Seeds of several orchid genotypes germinate better in a culture medium (Henrich, Stimart, and Ascher, 1981). Additions of banana pulp, coconut milk, or tomato juice to the medium stimulate the germination process (Northen, 1970).

Storage Conditions

Orchid seeds exhibit orthodox storage behavior. Seeds retain high viability and vigor under dry, cool conditions. Seeds lose their viability rapidly under room temperatures without desiccation. The percentage of germination is low in fresh seeds. Orchid seeds stored at 6°C showed high germination (90 percent) after ten weeks of storage, while at 45°C storage, they exhibited 40 percent seed viability in the presence of a desiccant. In the absence of a desiccant, the percentage of seed germination was reduced to 73 and 25 percent after ten weeks of storage at 6 and 45°C, respectively. Orchid seeds preserved below −70°C are supposed to retain 50 percent viability for a minimum period of two centuries (Thornhill and Koopowitz, 1992). Shoushtari et al. (1994) suggested storing orchid seeds at 4°C over calcium chloride to preserve viability for 10 to 20 years. Seed storage under cryogenic conditions maintains seed viability for longer periods (Koopowitz, 1986).

CACTUS

Introduction

Cacti include several genera and species used for ornamental purposes. Some of the important ones are *Astrophytum, Cephalocereus, Cerei, Cereus, Chamaecereus, Cleistocactus, Echinocereus, Echinopsis, Ferocactus, Lobivia, Nopalea, Opuntia, Rebutia,* and *Zygocactus.* Cacti are extremely hardy plants and can withstand dry conditions. They are grown in gardens, especially in rock gardens, or in pots for ornamental purposes. Plants are modified into various formations, which bear mostly short-lived flowers. Fruit is a fleshy, depressed globose. Certain cacti plants resemble particular animals, birds, and plants, giving them a unique appearance. Large genetic diversity in cacti occurs in South America. Seeds, cuttings, offsets, bulbils, and grafts are used for propagation.

Seed Storage

Seed Germination

Seed germination and seedling emergence is slow in cacti. Seeds of *Cereus peruvianus* Ricob germinate well at 15 to 25°C. The optimum temperature for germination of *Oreocereus trollii* seeds is between 15 and 20°C, and very few seeds germinate at 30°C (Zimmer, 1966). Seed germination is rapid and higher at alternate temperatures of 20/30°C in the presence of light (Ellis, Hong, and Roberts, 1985).

Storage Conditions

Cactus seeds appear to be orthodox in storage behavior. They preserve viability and vigor under dry, cool conditions. Seeds of *Cereus peruvianus* and *Oreocereus trollii* lose their viability rapidly under room temperatures and maintain it at 2°C (Zimmer, 1966). The viability of *Espostoa lanata* seeds decreases with an increase in storage period (Zimmer, 1967). Dry seeds are packed in polyethylene bags or in laminated aluminum foil pouches and stored at low temperatures for longer maintenance of seed viability.

GLADIOLUS (*Gladiolus* spp.)

Introduction

Gladiolus, or sword lily, is a perennial herbaceous plant with sword-shaped foliage. It is predominantly used as a cut flower and planted in beds or as a herbaceous border in a garden. Of the many species of *Gladiolus, G. colvillei, G. gandavensis, G. segetum,* and *G. tristis* are cultivated for ornamental purposes. Flowers are showy, of various colors, and arranged on one side of the spike. Flowers vary in shape, size, and color and have excellent keeping quality. The inflorescence contains up to 30 flowers, and colors include red, yellow, pink, violet, orange, and purple. Corms are commonly used for propagation. The spike length greatly depends on corm size. A medium-sized corm with a high crown gives better seedling growth during propagation than the large and flat corms. Gladioli are broadly grouped into large flowered, butterfly, and miniature types. The latter ones are suitable for bouquets and are grown in smaller gardens. Seeds are used in hybridization programs for the improvement of flower quality and the incorporation of resistance to various biotic and abiotic stresses.

Seed Storage

Seed Germination

Light and temperature influence seed germination and germination rate in gladioli. A constant temperature of 20°C is congenial for higher total germination (97 percent) (Carpenter, Wilfret, and Cornell, 1991), which takes about 16 days to complete (Ellis, Hong, and Roberts, 1985).

Storage Conditions

Gladiolus seeds are orthodox in storage behavior and their viability decreases at higher storage temperatures and relative humidity. Seeds remain viable for a period of one year under room temperatures. The germination percentage is reduced after 12 months of storage. The optimum storage conditions of 14°C and 26 percent RH are favorable for longer seed storage (Carpenter, Wilfret, and Cornell, 1991). Barton (1953) reported that high seed germination capability was retained for five days when seeds were stored in sealed containers at 5°C. Seed storage at –4°C extends the longevity to ten years or more. Seed moisture increases with repeated opening and closing of containers, thereby affecting seed viability. Thus, this should be avoided as much as possible during seed storage. However, reduction of seed moisture from 11.8 to 4.2 percent does not affect total germination.

AMARYLLIS: *Hippeastrum hybridum* Hort.

Introduction

Amaryllis is a popular, showy bulbous plant. The hybrids cultivated presently originated from crossing between *Hippeastrum reginae* and *H. vittatum*. These species are also grown in gardens for ornamental purposes. Amaryllis is planted in flower beds, borders, pots, window boxes, and used in landscaping and as cut flowers. Amaryllis grows up to 1 m in height and has strap-shaped leaves. The inflorescence is terminal and bears fragrant, trumpet-shaped flowers. Flowers form in a cluster of two to four and vary in color, from red, rose, white, and crimson, to orange and with red or white stripes. Fruit is globose and contains a few small green seeds. It is commonly propagated through the bulbs. Seeds are used for evolving suitable varieties through hybridization and for conservation of genetic diversity through seed storage.

Seed Storage

Seed Germination

Amaryllis seeds germinate better at alternate temperatures of 20/30°C for 16/8 h, respectively. Pindel (1990) indicated that a temperature of 20°C was optimum for germination and that a higher temperature affected seedling emergence. Amico and colleagues (1994) reported that storing seeds at 4°C for 30 or 60 days increased germination speed and seedling emergence but decreased the total germination. The percentage of germination is higher in large seeds than in small seeds. Seed germination improves with seed stratification. Germination takes 28 days to complete (Ellis, Hong, and Roberts, 1985).

Storage Conditions

Amaryllis seeds are fairly good storers under ambient conditions and show orthodox storage behavior. High seed moisture and high storage temperature reduce the storage life of seeds. Carpenter and Ostmark (1988) reported that seed viability is unaffected by the dehydration of seeds. Seed storage at lower temperatures of 2 to 6°C maintains their viability for eight months (Pindel, 1990). Seeds are tolerant to freezing, but viability decreases gradually as freezing temperatures decrease from 0 to −80°C. Seeds with low moisture content give a higher percentage of germination. Seeds having 11 percent moisture are viable for 12 months when stored at 5 or 15°C. Viability is affected by a higher storage temperature of 25°C or 35°C. Seed viability is lost completely within three months of storage at higher temperatures (25 to 35°C) and higher relative humidity (75 to 95 percent).

MARIGOLD (*Tagetes* spp.)

Introduction

Marigold is a popular garden flower that blooms for a long period (see Figure 43.1). It is also used as a cut flower and for decoration purposes due to its excellent keeping quality. Two types of marigolds, namely, African *(Tagetes erecta)* and French *(T. patula),* are commonly grown in gardens. The latter is dwarf in nature and bears profuse small flowers. This type is used in hanging baskets, window boxes, and rockeries. Its flowers are single or double and vary in shape, size, and color, such as yellow, orange, or red. It is a hardy plant that grows well in a wide range of climatic and soil conditions, and is commonly propagated from seeds. Occasionally, plants are also raised from cuttings. Furthermore, its seeds are mainly used in breeding programs for evolving new varieties and in the conservation of genetic diversity in the gene bank.

FIGURE 43.1. Marigold Plant in Bloom

Seed Storage

Seed Germination

Fresh seeds exhibit dormancy and require about three months of after-ripening before germination (Ramakrishnan, Khan, and Khan, 1970). Marigold seeds germinate better at a constant temperature of 20°C or at alternate temperatures of 20/30°C for 16/8 h, respectively, in the presence of light (Ellis, Hong, and Roberts, 1985). Seed germination is rapid and good after soaking seeds in potassium nitrate (0.75 percent) (Selvaraju and Selvaraj, 1994).

Storage Conditions

Marigold seeds exhibit orthodox storage behavior and store fairly well under ambient conditions. Seed drying is beneficial in maintaining high viability during storage. The lowering of seed moisture and seed storage at low temperatures preserve viability for a longer period. Nikolova-Khristeva (1973) reported that seed moisture greater than 9 percent was injurious to seeds during storage, causing them to lose viability rapidly. Seeds with 8 percent moisture showed 38 to 50 percent germination during the first year of storage, and 30 to 39 percent during the second year. Storing seeds with 5 to 6 percent moisture in moisture-proof containers at lower temperature is beneficial for longer seed storage. Seeds of cultivars Queen Sophia and Butter Scotch are stored in polyethylene (0.175 mm) and laminated aluminum foil pouches at 5 and −20°C. Seed viability is lost within a year under ambient conditions, whereas it is preserved for three years at 5 and −20°C storage (Doijode, unpublished data). However, Ramakrishnan, Khan, and Khan (1970) reported complete loss of viability by 21 months under ambient conditions.

DELPHINIUM (*Delphinium* spp.)

Introduction

Delphinium ajacis and *D. consolida* are two commonly grown *Delphinium* species that are planted in gardens and also used as cut flowers. They are commonly planted in borders or beds. Flowers are blue, purple, pink, or white and arranged in spikes. Delphinium is propagated through seeds. Seeds are also used in evolving new cultivars through hybridization and also in the conservation of genetic diversity in the gene bank.

Seed Storage

Seeds exhibit orthodox storage behavior and maintain high viability at low storage temperature and low relative humidity. Seed germination in delphinium is better at a constant temperature between 15 to 20°C or at alternate temperatures of 10/20°C or 15/25°C for 16/8 h. Delphinium seeds remain viable for a fairly long period under ambient conditions. Carpenter and Boucher (1992) reported that delphinium seeds could tolerate storage at low, nonfreezing, or subzero temperatures. The highest seed germination occurred following storage at 5°C and 30 to 50 percent RH. Air-dried seeds stored at 21.1°C satisfactorily preserve viability, but germination decreases after 38 months of storage (Crocker, 1930). Seeds gain moisture through frequent opening and closing of containers, thereby affecting seed viability. Seed viability preserves well for five years at 5°C and for ten years or more at –4°C (Barton, 1953).

PETUNIA: *Petunia hybrida* Hort.

Introduction

Petunia is a popular annual flower grown in gardens. Plants flower profusely and continue to bloom for a long period. The present-day cultivars are derived from *Petunia integrifolia* and *P. nyctaginiflora*. Petunia is excellent for flower beds, hanging baskets, pots, and window boxes. Flowers are showy, solitary, axillary, and funnel shaped; they are white, yellow, bluish, pink, and red in color and fragrant. Fruit is a capsule containing numerous small seeds. Seeds are commonly used for propagation. They are also used in evolving new cultivars through hybridization and in the conservation of genetic diversity in the gene bank.

Seed Storage

Seed Germination

Petunia seeds germinate better at alternate temperatures of 20/30°C for 16/8 h in the presence of florescent light (Ellis, Hong, and Roberts, 1985). Corbineau and Come (1991) reported that petunia seeds germinate at between 15 to 35°C and the optimum temperature is 30°C. Jassey and colleagues (1977) studied germination behavior in 25 lines of *Petunia* species and hybrids up to 50 weeks after harvest. Though seeds of a few lines germinated without any dormancy throughout storage, seeds of a few species did not germinate at 25 or 30°C unless treated with gibberellic acids or exposed to light. Seeds of *P. parodii* remained dormant even after 50 weeks of storage, as they contain inhibitors in the testae and germinate readily on isolation of embryos and treatment with gibberellic acid.

Storage Conditions

Petunia seeds survive for a short period under ambient conditions, and they exhibit orthodox storage behavior. High seed viability is maintained under dry, cool conditions. Seeds stored in open boxes or paper packets lose viability considerably within four months and completely by nine months of storage under ambient conditions (3 to 30°C and 32 to 72 percent RH). Seed viability is preserved for three to four years in sealed plastic bags and also by additions of silica gel (Lowig, 1969). Seeds stored in sealed containers under refrigerated conditions (5 to 7°C) retain viability for five years (Baigonina and Zorina, 1984).

GERBERA: *Gerbera jamesonii* Adlam

Introduction

Gerbera is largely grown for its cut flowers but is also planted in rock gardens, flower beds, borders, and pots. It is a dwarf perennial herbaceous plant that grows up to 45 cm in height. Leaves form a cluster at the base. Flowers are single or double with perfect disc and ray flowers that can be yellow, salmon, pink, red, white, or orange (see Figure 43.2). Flowers have a vase life of 10 to 15 days. Gerbera prefers cool conditions for optimum growth. Seeds and division of clumps are used in propagation.

Seed Storage

Seed Germination

Gerbera seeds germinate at a constant temperature between 15 and 25°C (Carpenter, Ostmark, and Cornell, 1995). However, Ellis, Hong, and Roberts (1985) reported that a constant temperature of 20°C or alternate temperatures of 20/30°C for 16/8 h are ideal for rapid and higher seed germination. A longer germination period results from alternating temperatures. Seed germination is affected by reduction of moisture below 3.5 percent (Carpenter, Ostmark, and Cornell, 1995).

Storage Conditions

Gerbera seeds are short-lived under ambient conditions, and they show orthodox storage behavior. Seeds preserve high viability and vigor for a

FIGURE 43.2. Gerberas in the Flowering Stage

longer period under dry, cool conditions. Carpenter, Ostmark, and Cornell (1995) reported that moisture reduction from 7.1 to 3.5 percent had no effect on germination. Seeds lose viability after nine months of storage under ambient conditions but preserve well for 30 months at –5 to 5°C (Franceschetti, 1973). Carpenter, Ostmark, and Cornell (1995) observed similar germination percentage after storage at temperatures between 5 and –5°C, whereas germination was reduced when seeds were stored at –10°C or below. Seed viability can be preserved for 12 months at storage conditions of 5°C and 32 percent RH. Seeds stored at moderate relative humidity (52 percent) showed low viability, and none of them germinated when stored at high relative humidity (75 percent) and high temperature (25°C) for 12 months.

PERIWINKLE: *Catharanthus roseus* (L.) Don

Introduction

Periwinkle is grown in pots, rockeries, window boxes, and as a ground cover. Plants are bushy and dwarf in habit, growing up to 60 cm in height. Leaves are oval and shiny. Plants flower profusely for a long time. Flowers are round or flat and white or pink in color. Seeds and cuttings are used for propagation. Seeds are also used in crop improvement through hybridization and also in the conservation of genetic diversity in the gene bank.

Seed Storage

Periwinkle seeds are small, 2 to 3 mm in length and 1 to 2 mm in width. One thousand seeds weigh about 1 to 2 g. Fresh seeds readily germinate and exhibit 95 to 96 percent germination. Seed viability becomes zero after two years of storage under improper storage conditions. However, Mitrev (1976) reported that seed viability decreased gradually under ambient conditions. The germination percentage was 95, 53, 28, and 0 after one, two, three, and four years of storage, respectively, under ambient conditions. Seeds stored in desiccators over calcium chloride showed 49 percent germination after four years of storage (Gogitidze, 1983). Periwinkle seeds retain viability for a fairly long period under ambient conditions and exhibit orthodox storage behavior. For longer seed storage, low-moisture seeds are packed in thick polyethylene bags or laminated aluminum foil pouches and stored at low temperatures.

RHODODENDRON (*Rhododendron* spp.)

Introduction

Rhododendron is an evergreen or deciduous small perennial shrub grown in temperate zones for its beautiful flowers. Several *Rhododendron* species are grown in gardens for ornamental purposes, such as *R. indicum, R. japonicum, R. kotschyi, R. luteum, R. macgregoriae, R. obtusum, R. occidentale, R. simsii, R. sinense, R. vaseyi,* and *R. viscosum*. Flowers are borne terminally in clusters; they are large, single or double, and vary in color, such as pink, red, scarlet or carmine, and white. Seeds, cuttings, and grafts are used in propagation.

Seed Storage

Seed Germination

Rhododendron seeds exhibit dormancy. Fresh seeds of *R. kotschyi* showed 6 to 12 percent germination, reaching a maximum of 30 to 43 percent in 8 to 11 months of storage (Cherevko and Sapozhenkova, 1975). The optimum temperature for germination is 20°C. Lee, Song, and Hong (1982) reported that seed washing with water adversely affected germination, whereas germination improved with seed soaking in thiourea (1 percent) and indolebutyric acid (100 ppm).

Storage Conditions

Rhododendron seed viability is preserved for a short period under ambient conditions, particularly a few weeks at room temperatures. Seeds show orthodox storage behavior. Furthermore, deterioration decreases at lower storage temperatures and relative humidity. Germination of seeds of *R. macgregoriae* was reduced by 10 percent in two weeks when seeds were stored at 27°C and in 85 weeks at 4°C and less than 5 percent RH. Short seed life is not due to recalcitrance. In addition, such seeds dry well for longer storage at lower temperatures (Rouse and Williams, 1986). Fillmore (1949) recommends storing rhododendron seeds in the refrigerator in sealed containers. Seeds of *R. kotschyi* preserved for four to five years show low germination percentage (Cherevko and Sapozhenkova, 1975).

ANTHURIUM (*Anthurium* spp.)

Introduction

Anthurium is grown for its attractive foliage and beautiful, showy flowers. The *Anthurium* species most commonly grown for their foliage are *A. clarinervium, A. crystallinum, A. digitatum, A. magnificum, A. veitchii,* and *A. warocqueanum,* and the species commonly grown for their flowers are *A. andraeanum* and *A. scherzerianum.* Anthurium grows well in shade and in warm, humid conditions. It is also grown in pots for indoor gardening. Plants are dwarf and compact, with large, dark green, velvety, heart-shaped or ovate leaves. The inflorescence consists of a spathe, and the spadix is showy and covered with bisexual flowers. Fruits are striking and brilliantly colored. Seeds, suckers, and stem cuttings are commonly used in propagation. Seeds are also used in hybridization programs for crop improvement.

Seed Storage

Seed Germination

Anthurium seeds show low viability and germinate poorly under in vivo conditions. Embryos are small and poorly developed, causing delayed and poor germination. The excised embryos germinate better under in vitro conditions.

Storage Conditions

Anthurium seeds exhibit recalcitrant storage behavior (Stanwood, 1987). They survive for an exceptionally short period under ambient conditions.

Seeds lose viability rapidly on desiccation, and it is difficult to store them longer, even under moist conditions; furthermore, high seed moisture is favorable for the growth of pathogens. However, seed viability is preserved for a shorter period when seeds are stored in berries. Bachthaler (1979) reported that berries disinfected with benomyl, zineb, captan, and thiram maintained viability (60 percent) for 16 weeks at 10°C, whereas berries stored at 2°C suffered cold injury, and those stored at 5°C were affected by fungal infections and cold injury. Seed storage in a low-pressure cabinet (700 and 400 hectopascals [hPa]) preserved viability for 20 weeks at 11°C and gave 70 and 50 percent germination, respectively (Bachthaler, 1993).

CALADIUM: *Caladium bicolor* (Ait.) Vent.

Introduction

Caladiums are predominantly grown in pots or in garden beds for their beautiful foliage. Leaves are broad and of various colors, such as red, rose, salmon, white, green, pink, and purple, in a speckled pattern. Seeds and bulbs are used in propagation. Seeds are also used in hybridization programs for evolving new genotypes and in the conservation of genetic diversity.

Seed Storage

Caladium seeds germinate immediately after harvest, and they require continuous light for germination. Seed germination decreases from 95 to 87 percent when moisture content falls below 14 percent. These seeds appear to be recalcitrant in storage behavior (Stanwood, 1987) and maintain high viability under cooler conditions. Caladium seeds retain high viability for seven days under a wide range of low temperatures (10 to –80°C) without loss of viability. Further, seeds are successfully preserved for six months at 15°C and 22 percent RH, without a change in initial viability. However, seed viability will be significantly reduced at a higher temperature (25°C) and higher relative humidity (95 percent) (Carpenter, 1990).

GERANIUM (*Pelargonium* spp.)

Introduction

Geranium is grown for its beautiful foliage and flowers. The commonly grown geraniums are *Pelargonium peltatum, P. radens,* and *P. zonale.* Leaves are dark green and round. Flowers are round, single or double, and come in various colors, such as pink, scarlet, crimson, red, and purple. Flowers bloom over a long period. Geranium is suitable for planting in pots, beds, or window boxes. Seeds and terminal cuttings are used in propagation.

Seed Storage

Geranium seeds germinate better at alternate temperatures of 20/30°C for 16/8 h or at a constant temperature of 20°C, and germination takes 28 days to complete (Ellis, Hong, and Roberts, 1985). Geranium seeds store fairly well under ambient conditions. They show orthodox storage behavior and deteriorate rapidly at higher storage temperatures and relative humidity. Seeds remain viable for a long period under cool, dry conditions. They can be preserved for five years when stored at 20°C, showing 100 percent germination (Bachthaler, 1983).

CAMELLIA (Camellia spp.)

Introduction

Camellias are popular ornamental shrubs. Many of the existing varieties are hybrids of *Camellia japonica* and *C. saluenensis*. The plant is an evergreen, and the flowers are attractive, bisexual, single, axillary or terminal, funnel shaped, and white or pink. Camellia is commonly propagated by stem cuttings and grafting. Seeds are used for evolving new genotypes through hybridization.

Seed Storage

Camellia seeds appear to be orthodox in storage behavior and retain high viability under cool, dry conditions. High seed viability can be preserved at low temperatures and low relative humidity. Seeds stored at 4.4°C maintain viability for one year (Griffith, 1955). Freshly collected seeds should be kept for 18 days at room temperatures, then covered with plastic film and stored at 0 to 2°C to maintain high germination percentage for more than two years (Han, 1984).

REFERENCES

Amico, R.U., Zizzo, G.V., Agnello, S., Sciortino, A., and Iapichino, G. 1994. Effect of seed storage and seed size on germination, emergence and bulblet production of *Amaryllis belladonna*. *Acta Hort*. 362:281-288.

Bachthaler, E. 1979. Studies on the storage of seeds of *Anthurium scherzerianum* hybrids. *Die Gartenbauwissenschaft* 44:251-255.

Bachthaler, E. 1983. Pelargoniums: Scarified seed experiments provide information. *Skaritiziarte Saat Versuche geban Aufschlusse* 83:1103.

Bachthaler, E. 1993. Studies on the storage of seeds of *Anthurium scherzerianum* hybrids at reduced pressure. *Die Gartenbauwissenschaft* 58:21-24.

Baigonina, V.P. and Zorina, M.S. 1984. Seed viability of annuals in relation to storage length. *Refrative Zhurnal* 2.55.882:55.

Barton, L.V. 1953. Seed storage and viability. *Boyce Thompson Inst. Contrib.* 17:87-103.

Bass, L.N. 1980. Flower seed storage. *Seed Sci. Technol.* 8:591-599.

Benedetto, A.H. and Tognetti, J.A. 1985. Effects of commercial storage on the germination of seeds of annual ornamental species. *IDIA* No. 441:112-117.

Blundell, J.B. and Jackson, G.A.D. 1971. Rose seed germination in relation to stock production. *Rose Ann. London,* pp. 129-135.

Carpenter, W.J. 1990. Light and temperature govern germination and storage of caladium seeds. *HortSci.* 25:71-74.

Carpenter, W.J. and Boucher, J.F. 1992. Temperature requirements for the storage and germination of *Delphinium cultorum* seeds. *HortSci.* 27:989-992.

Carpenter, W.J. and Ostmark, E.R. 1988. Moisture content, freezing and storage conditions influence germination of amaryllis seed. *HortSci.* 23:1072-1074.

Carpenter, W.J., Ostmark, E.R., and Cornell, J.A. 1995. Temperature and seed moisture govern germination and storage of gerbera seeds. *HortSci.* 30:98-101.

Carpenter, W.J., Wilfret, G.J., and Cornell, J.A. 1991. Temperature and relative humidity govern germination and storage of gladiolus seeds. *HortSci.* 26:1054-1057.

Cherevko, M.V. and Sapozhenkova, T.V. 1975. Seed vigor and emergence in *Rhododendron kotschyi. Ukrainskii Bot. Zhurnal* 32:361-366.

Corbineau, F. and Come, D. 1991. Seeds of ornamental plants and their storage. *Acta Hort.* 298:313-321.

Crocker, W. 1930. Harvesting, storage and stratification of seeds in relation to nursery practices. *Florists' Rev.* 65:43-46.

Crocker, W. and Barton, L.V. 1931. After-ripening, germination and storage of certain rosaceous seeds. *Boyce Thompson Inst. Contrib.* 3:385-404.

Cromarty, A.S., Ellis, R.H., and Roberts, E.H. 1982. *The design of seed storage facilities for genetic conservation.* Rome: International Board for Plant Genetic Resources, pp. 1-96.

Ellis, R.H., Hong, T.D., and Roberts, E.H. 1985. *Handbook of seed technology for gene banks,* Volume II. Rome: International Board for Plant Genetic Resources, pp. 1-667.

Fillmore, R.H. 1949. Growing rhododendron from seeds. *Arnoldia* 9:45-51.

Foster, T.C. and Wright, C.J. 1983. Germination of *Rosa dumetorum* Laxa. *Scientific Hort.* 34:116-125.

Franceschetti, U. 1973. Observations on the longevity of gerbera seeds. *Sementi Elette* 19:7-10.

Frei, J.K., Fodor, R.C., and Haynick, J.L. 1975. The suitability of certain barks in growth media for orchids. *Am. Soc. Orchids Bull.* 44:51-54.

Gogitidze, Ts. 1983. Characteristics of *Catharanthus roseus* seed biology. *Subtrop. Kul.* No. 6:102-104.

Griffith, A. 1955. Culture and classification of *Camellia* and related genera. *Fla. Agr. Exp. Stn. Ann. Rep.*, p. 111.

Grzesik, M. 1993. Effect of storage conditions on the viability of ornamental seeds. *Proc. Fourth Inter. Workshops on Seeds Angers, France* 3:799-803.

Haenchen, E. 1968. Problems of stratification of rose rootstock seeds. *Die Gartenbauwissenschaft* 15:270-272.

Han, N.L. 1984. Studies on the technique of low-temperature storage of seeds of *Camellia oleifera. Forest Sci. Technol.* No. 12:7-9.

Harrington, J.F. 1968. Moisture equilibrium values for several grass and legume seeds. *Agron. J.* 60:594-597.

Henrich, J.E., Stimart, D.P., and Ascher, P.D. 1981. Terrestrial orchid seed germination in vitro on a defined medium. *J. Am. Soc. Hort. Sci.* 106:193-196.

Hossain, M.Z., Mannan, M.A., and Roy, S.K. 1996. Effect of seed age on germination and protocorm development of an orchid *Vanda roxburghi* in vitro. *2nd Inter. Crop Sci. Congr. New Delhi* Abstr. No. P14-005:347.

Jassey, Y., Monin, J., Cornu, A., and Dommergues, P. 1977. Preliminary analysis of the variance in the states of seed dormancy in several petunia lines. *Comptes Rendus Hebdomadaris des Seances de l'Academic des Sci.* 28:1797-1800.

Koopowitz, H. 1986. A gene bank to conserve orchids. *Am. Orchid Soc. Bull.* 55:247-250.

Lee, G.E., Song, Y.N., and Hong, H.O. 1982. Studies on the wild *Rhododendron fauriei* from rufescens in Korea. *J. Korean Soc. Hort. Sci.* 23:64-69.

Linden, B. 1980. Aseptic germination of seeds of northern terrestrial orchids. *Ann. Bot. Fen.* 17:174-182.

Lowig, E. 1969. The storage property of petunia seeds. *Saatgut Wirtsch.* 21:28.

Mckelvie, A.D. and Walker, K.C. 1975. Germination of hybrid tea rose seeds. *J. Hort. Sci.* 50:179-181.

Mitrev, A. 1976. Laboratory investigation on the biological properties of *Vinca rosea* seeds produced in Bulgaria. *Rasten. Nauki.* 13:89-90.

Nikolova-Khristeva, N. 1973. Changes in the moisture contents and germination of stored seeds of some annual flowers. *Nauchni·Trudove Vissh Sel Inst. Vasil Kolarov* 22:43-48.

Northen, R.T. 1970. *Home orchid growing.* New York: Van Nostrand Reinhold Company.

Pindel, Z. 1990. Effect of conditions and time of seed storage on the germinability in hippeastrum. *Prace Instytin-Sadowinct* 15:51-58.

Popcov, A.V. and Buc, T.G. 1966. Presowing treatments of seeds of the dog rose *Rosa canina* L. *Bull. Glav. Bot. Sada* No. 62:30-34.

Ramakrishnan, V., Khan, W.M.A., and Khan, A.M.M. 1970. Studies on the germination of seeds of a few ornamental flowering annuals. *South Indian Hort.* 18:93-95.

Rouse, J.L. and Williams, E.G. 1986. Storage life of Vireya rhododendron seed as affected by temperature and relative humidity. *Seed Sci. Technol.* 14:669-674.

Selvaraju, P. and Selvaraj, J.A. 1994. Effect of presowing treatments on germination and vigor of seeds in marigold *(Tagetes erecta). Madras Agril. J.* 81:496-497.

Serebryakova, N.V. and Kalanova, A.I. 1977. The effect of water soluble vitamins on rose seed germination and rooting of cuttings. *Vitamins Rasti Resu Ispo Moscow,* pp. 215-221.

Shoushtari, B.D., Heydari, R., Johnson, G.L., and Arditti, J. 1994. Germination and viability staining of orchid seed following prolonged storage. *Lindleyana* 9:77-84.

Smith, S.E. 1974. Amybiotic germination of orchid seeds on carbohydrates of fungal origin. *New Phytology* 72:497-499.

Stanwood, P.C. 1987. Storage and viability of ornamental plant seeds. *Acta Hort.* 202:49-56.

Thornhill, A. and Koopowitz, H. 1992. Viability of *Disa uniflora* seeds under variable storage conditions: Is orchid gene banking possible? *Biol. Conservation* 62:21-27.

Yu, T.Y., Yeam, D.Y., and Kim, Y.J. 1975. The level of gibberellic acid and abscisic acid-like substances in *Rosa multiflora* seeds as affected by low-temperature treatments. *J. Korean Soc. Hort. Sci.* 16:114-119.

Zimmer, K. 1966. Investigations on the influence of temperature on the germination of cactus seeds. IV. The germination of *Cercus peruviana* and *Oreocereus trolli. Die Gartenbauwissenschaft* 31:437-445.

Zimmer, K. 1967. Investigations on the influences of temperature on the germination of cactus seeds. V. The germination of *Espostoa lanata* and *Echino cactusgrusoni. Die Gartenbauwissenschaft* 32:173-179.

Index

Page numbers followed by the letter "f" indicate figures; those followed by the letter "t" indicate tables.

T - #0477 - 101024 - C0 - 212/152/19 - PB - 9781560229018 - Gloss Lamination